Learning To Live Together: Promoting Social Harmony

J. A. Scott Kelso
Editor

Learning To Live Together: Promoting Social Harmony

 Springer

Editor
J. A. Scott Kelso
Human Brain and Behavior Laboratory,
   Center for Complex Systems and Brain
   Sciences
Florida Atlantic University
Boca Raton, FL
USA

and

The Intelligent Systems Research Centre,
   Ulster University (Magee Campus)
Derry ~ Londonderry
Northern Ireland

ISBN 978-3-319-90658-4   ISBN 978-3-319-90659-1   (eBook)
https://doi.org/10.1007/978-3-319-90659-1

Library of Congress Control Number: 2018940637

Printed on acid-free paper

This Springer imprint is published by the registered company Springer International Publishing AG
part of Springer Nature
The registered company address is: Gewerbestrasse 11, 6330 Cham, Switzerland

This book is dedicated to the life and work of Epimenidis Haidemenakis, the founding father and creator of the idea of The Olympiads of the Mind. Epimenidis was born on May 4, 1932 in Chania, Crete, Greece. As the youngest son of a Brigadier General in the Greek Army and Valedictorian of his Gymnasium in Chania, Crete, Epimenidis, or Epi as he is affectionately known to his colleagues throughout the world, has always held close to his heart the classic Greek philosophy—that competition is not about competing "against" but about competing to better oneself and one's fellow competitors.

Epi came to the United States on a scholarship to Columbia University in New York where he earned his BA degree in Physics in 1954. After completing his undergraduate studies at Columbia, Epi did research in solid state physics at various laboratories in the US including MIT National Magnets Lab, Honeywell Research Center, RCA Engineering Laboratory and the U.S. Naval Research Laboratory. His research took him to France (L'Ecole

Normale Superieure) where he continued his studies at the University of Paris.

After 12 years of scientific research experience Epi realized that he might be more skillful and productive in science administration and so continued his career as a consultant in international science, technology, industry and educational policy, conducting studies and ex- post evaluations for corporations, governments and intergovernmental organizations. These included International Telephone and Telegraph, International Harvester, Alsthom-Atlantique, the Ministry of Industry of France, UNESCO, UNIDO, UNDTC, and the Organization for Economic Cooperation and Development. On returning for the first time to his hometown of Chania, Epi felt that the students needed to connect with the giants in science and technology and that there was a need to offer new knowledge to these young students. So he invited his friends from a Physics Conference on Semiconductors in which he participated in Kyoto in 1966 to organize a conference in Crete. Epi managed to bring together well known physicists and the 1st Conference on the Physics of Solids in Intense Magnetic Fields took place in July 1967.

It was such a great success and the reception to Epi's idea of further international scientific interchange was so overwhelming, that it sparked the first of nineteen "Chania Scientific Conferences" mostly on topics in Physics. The success of these Conferences inspired Epi to form the International Science

Foundation—a world renowned organization to this day. He then pursued a Ph.D. in International Relations which was awarded to him in 2008 from Camden University.

The International Science Foundation developed a stellar reputation in the 1970's and 80's inviting Nobel Laureates from all over the globe. In 1991, the scope was broadened to include all disciplines and the first Olympiad of the Mind was held under the newly formed International S.T.E.P.S. Foundation (Science, Technology, Economics and Politics for Society), a U.S. non-profit organization. By definition, S.T.E.P.S. begins with the inherent and inevitable human curiosity for scientific research. In turn, technology, being the application of science, has a direct impact on the economy, which itself influences the flow of politics to a large extent.

In unison, these four elements are of major importance in affecting our lives, and thus have distinct consequences for our society. The Foundation, by bringing together people belonging to countries with different or competing national interests, worldviews and perceptions of global peace, created a context of mutual understanding and trust. Many friendships were developed, even between nationalities which were in political and ideological conflict then (such as American–Russian, vodka-whiskey exchanges). As the intellectual counterpart to the Olympic Games for the body, Epi's Olympiads of the Mind (OM) assembled some of the world's most powerful minds to apply the synergy of

Science, Technology, Economics and Politics to the major global and regional challenges facing our Society.

The ultimate goal of each Olympiad of the Mind is to contribute to the development of humankind by formulating policy recommendations for action and solutions to problems of global urgency. Throughout the years, Epi has ensured that the Olympiads remain free of any political agenda, controlling body or influence, so that uncensored and "real" solutions can be proposed for world problems. For example, the Olympiad held after the 9–11 World Trade Tower attack in New York was at the United Nations Educational, Scientific and Cultural Organization (UNESCO) in Paris in 2005 with the aim of investigating ways to apply communication technologies and information systems to solve the global challenges of terrorism, poverty, unemployment, nuclear proliferation and energy resources.

It has not been the role of the International S. T.E.P.S. Foundation (STEPS) to implement the findings of each OM. Instead, STEPS has sought to stimulate and invite the most powerful international organizations, like the World Bank and the United Nations, to do so.

In this respect, and in the spirit of its founder, Epimenidis Haidemenakis, we are currently seeking more ways to implement these recommendations (see Section VI). We invite our readers and seek their input to find channels to share and actively engage this information in their sphere of influence however large or small that may be.

To our friend and colleague Epimenidis Haidemenakis, a great visionary and true humanitarian, a person devoted to bringing great minds together to create solutions for The Peoples of the World to Learn to Live Together, we dedicate this book in awe of the trail that he has blazed so brightly. We pledge ourselves to carry on the work of the Olympiads of the Mind with the spirit of excellence in diversity, variability and compassion that Epi initiated and has so aptly demonstrated throughout his life.

Konny Light, JD, Corporate Secretary
J. A. Scott Kelso, Ph.D. Hon. MRIA,
Chairman and President
International S.T.E.P.S. Foundation and the
Olympiads of the Mind

Preface

This volume consists of a number of papers presented at the Eighth and Ninth Olympiads of the Mind (OM) hosted by the US National Academies of Science and Engineering in Washington, DC, in November 2007 and at the Orthodox Academy of Crete in Chania, Crete, in September 2017. Although separated in time by a period of 10 years, the two meetings shared a common goal. On the one hand, the title of the Washington meeting was "Brain Research: Improving Global Harmony"; on the other hand, the Crete meeting was called "Learning to Live Together." The present volume "Learning To Live Together: Promoting Social Harmony" thus represents a merger of both themes.

On such matters as global harmony and learning to live together, separation in time does not mean "outdated." As is quite clear from current events, the issue of how we can learn to live together in the face of division and conflict is not going away anytime soon. It has, one might say, a certain staying power. Though it lies at the very heart of our existence, humans, it seems, have not yet found a way to deal with it. What hope, then, that we humans can learn to live together with machines—or they with us?

Both the 8th and 9th Olympiads featured distinguished international scholars, including government and corporate representatives, leading researchers and academics from multiple disciplines, and Nobel Laureates—all committed to understanding the role and impact of science, technology, economics, and politics on human beings and society at large, with the aim of improving the human condition and achieving greater cooperation among people. The six sections of the current volume embrace issues of the environment, sustainability, and security; of diversity and how to achieve integration and peace among people in a fractured world; of healing and understanding ourselves, including the important role of brain research; of how to overcome poverty and inequality and how to improve education at all levels; and of how new technologies and tools can be used for common benefit, their drawbacks, and potential uses, from the smallest scales in nature to how we conduct financial transactions. The culmination of the book is a call to action, to join what one might call the "OM Movement"—bringing the best minds in the world together to create solutions to world issues so that we can all live together in harmony.

It must be said that the Olympiads of the Mind are a testament to the kind of creative richness and coordination that can emerge when heterogeneity (of people, the fields they represent, and the places they come from) is combined with open, engaged exchange. They are a forceful demonstration of the processes that, at least under some circumstances, enable human beings from very different backgrounds and cultures to work together to move us forward. Much more about the mission, vision, and history of the OMs is to be found in the dedication of this book to the extraordinary human being who had the insight and the drive to make them happen, Epimenidis Haidemenakis. His are difficult shoes to fill.

It remains now to thank all the contributors to this volume and all the staff and volunteers in Washington, DC, and Chania, Crete, who made these "meetings of the minds" not only possible, but such an enjoyable experience. The setup and smooth running of events relied on the skills and dedication of folks like Dr. Jill Harrington in Washington; Calli Cavros, Manolis (Lakis) Planas, Yannis Marinakis, Mirela Manos, and Leah Petroski in Crete; and Amruta Dadhe in Atlanta, Georgia (USA). IT support was gratefully provided by Vasileios Mavrikios. On behalf of The Olympiads of the Mind and the S.T.E.P.S Foundation, we thank you with all our heart.

I am very grateful to my physicist colleagues at The Center for Complex Systems and Brain Sciences, Armin Fuchs and Roxana Stefanescu, for helping organize the varied, individualistic styles and contents of the papers into a (relatively) common format; to Jane Buck for her willingness to edit one or two of the papers; and finally, to Tom Ditzinger of Springer, who immediately embraced the idea of a book project, harmonized the individual chapters, and sped up the production process, I am forever grateful. Dear Epi, this book is for you.

Boca Raton, USA   J. A. Scott Kelso
Derry, UK
January 2018

Contents

Part V   Technology

Part VI   Messages and Recommendations

Dr. Epimenidis Haidemenakis' Speech for the 9th Olympiad of the Mind

I am delighted to welcome you all here to the 9th Olympiad of the Mind (OM), with the theme of "Learning to Live Together," and thrilled to see such a special group of speakers and audience in attendance. I would like to welcome His Eminence Amphilochios, Bishop of Kissamos and Selinon, Representative of the Ecumenical Patriarchate of Constantinople as well as Dr. Kostas Zorbas, General Director of the Orthodox Academy of Crete. I would also like to make a reference to the late Archbishop Eirinaios, who created this conference centre and who was a good friend and a doer, not just a listener.

I represent the S.T.E.P.S. Foundation, which is a not-for-profit and tax-exempt educational and scientific corporation. Its function is to organize international forums dealing with the hard sciences, the life sciences, and the behavioral sciences.

Central to the philosophy of the OM is the concept that relies on the interdependence and synergy of five individual pillars of life, namely: science, technology, economics, politics, and society. By definition, S.T.E.P.S. begins with the inherent and inevitable human curiosity for scientific research. In turn, technology, or the application of science, has a direct impact on the economy, which itself to a great extent influences the flow of politics. In unison, these five elements are of paramount importance in affecting our lives and thus have distinct consequences on our society.

Tragic events such as the terrorist attacks on September 11, 2001, have shown to the world how the inequalities between industrialized and developing nations are not without serious consequence. While the civilization we now live in might rightly be described as the Communication Civilization, we also see an upsurge in war, injustice, and an ever-increasing gap between developed and developing nations. International relations are issues of security, energy resources, and territorial claims. The aggressive methods used to combat such problems are often counterproductive and expensive. Moreover, at the heart of many of the challenges faced by the world lies a lack of communication and understanding. The great technological advances of our Communication Civilization are not being optimally applied to solving its problems.

From the moment life appeared on this planet, humanity has tried to create a functional and humane society in which people can live together in harmony. But all those billions of years have gone by, and with today's easy solutions and convenient life, we still have not found a formula to live by peacefully. Learning to live together seems to be a very difficult thing, if not impossible. Humanity is trying to better itself and learn to live together, but during this process, we seem to have lost focus. Instead of learning to live in peace, we are learning how one nation can dominate the other.

That thought led me to our theme "learning to live together" (LTLT). That being said, I cannot hide from you that I feel more pessimistic than any other time we have discussed this. While the civilization we now live in might rightly be described as the Communication Civilization, we also see an upsurge in war, injustice, and the ever-increasing gap between developed and developing nations. Added to that, terrorism has increased through the last few years, and instead of learning to live together, we seem to be learning how to kill each other.

World War II is a vivid example of that. During that terrible war, the Soviet Union mourned on average nine dead every minute: every hour, 507 dead, and every day, 1400 dead. Unimaginable material destruction took place as well. Over 1700 cities, over 70,000 villages, thousands of hospitals, schools, libraries turned to dust on the ground of the USSR by the Nazis. And the killings go on. The casualties in the Japanese cities, as a result of the atomic bombings, were grave. Within the first two to four months following the bombings, the acute effects of it had killed about 146,000 people in Hiroshima and 80,000 in Nagasaki. Roughly, half of the deaths in each city occurred on the first day. The third example of humanity learning only to kill each other is the terror happening at our doorstep, in some Mediterranean countries, Iran, North Korea, and so on.

But we must fight this endless war with our best weapons, knowledge, ideas, and communication. Communication is vital to human beings; without it, a full and rich life, both exterior and interior, is unimaginable. It allows us to learn from others and work powerfully in teams, which is the only way people can actually learn to live in peace.

In terms of ideas, I was thinking about what could my contribution be in this OM, and I finally found it! It is women! If we manage to bring women's presence more into politics and seriously increase the number of women who possess political power, then we would be improving life and creating a more peaceful world.

Women give birth to life and have a natural inclination to protect and care for others. Therefore, they do not consider war as a possible solution, since by definition war is mainly destruction. Also, using the same argument, women would support the idea of formulating new legislation that protects the weak and poor and brings equality into existence. In addition, they are more likely to reason and be willing to explore compromise as well as seek and respect other people's opinions.

If capable women get placed in critical policy-making positions, they would be able to improve life as we know it and balance some dominating and aggressive aspects of men.

Throughout the years, men have shown that they are not capable of preserving life or improving it. Even in the world we live in today, with colossal technological breakthroughs and solutions that are supposed to make our lives easier, our leaders seem to not grasp the apparent fact that war should not be the solution to any problem and that every single war is taking us centuries back.

Thus, learning to live together seems an impossibility to me at times. Unless our political leaders take new kinds of measures, we will all end up with that feeling of pessimism that I know most of us have.

But, this is not the only time that mankind has faced such fragile situations, so, I hope, by the 10th OM, by using diplomacy and smart thinking, the bloodshed will be reduced and we will be looking at the future with more hope. And we will have created an International Committee of the Olympiads of the Mind (ICOM).

I will finish my speech with a nice phrase that we learned in high school in Latin class, "Dum Spiro, Spero" = "While I breathe, I hope."

Ladies and Gentlemen,

I thank you.

Part I
The State of the World

Global Response to Global Problem

Yuan T. Lee

In 1982 when the Erice Statement was written to address the real danger of nuclear war, the fear was that technology would be used not by the culture of love, but the culture of hatred to kill and destroy. These dangers of mass destructions are still very much with us today with more countries and groups trying to acquire nuclear weapons.

However, I would argue that the greatest danger we face today is of a different kind. I'm talking, of course, about climate change. The Erice Statement speaks of mankind having, for the first time in history, the military power to destroy the world. But we also, for the first time, have the power to change our environment to the point where it cannot support life anymore. And that power is not only military. With climate change, it is not enough that we stop using technology for *the culture of hatred*. Sometimes, technology for *the culture of love* can be a part of the problem.

I would like to take us back in time, to the beginning of the story of human development. I wonder how many people today can remember a time, when nobody could question that humanity was a part of nature. We used to depend on the Sun for almost everything. Photosynthesis converted solar energy and made so much possible: food to nourish us; materials to build shelter; clothes to warm us. Mankind had a relatively small footprint on the environment. Actually, I experienced this kind of lifestyle 72 years ago. I was just a boy, about 8 years old. World War II had spread to Taiwan, so I went to live in the mountains to avoid the danger of daily bombing of the cities by US airplane. We depended almost completely on the

The author is former President of Academia Sinica, Taiwan. He was awarded the Nobel Prize in Chemistry in 1986 (along with Dudley Herschbach and John Polanyi) for his work on understanding the dynamics of chemical reactions.

Y. T. Lee (✉)
Academia Sinica, Taipei, Taiwan
e-mail: ytlee@gate.sinica.edu.tw

© Springer International Publishing AG, part of Springer Nature 2019
J. A. S. Kelso (ed.), *Learning To Live Together: Promoting Social Harmony*,
https://doi.org/10.1007/978-3-319-90659-1_1

nature around us. And to this day, I remember it as one of the happiest times of my life.

But starting about 250 years ago, something happened. The industrial revolution began in England, and then spread to Germany, France, and the United States. We also discovered that there were these black stuff in the ground that contained huge amounts of energy. And everything changed. It seemed like humanity was breaking free of nature's limits.

Our technology advanced rapidly. It began with the steam engine, the weaving machine, and the internal combustion engine, and continued with the elevator, telegraph, atomic energy, etc. Advances in chemistry allowed us to synthesize materials that never existed before: nylon, polyester, carbon fiber… we invented robots and sent them deep into the oceans and into space. By exploiting fossil fuels, we dissociated ourselves from dependence on the sun. Now, human development was no longer limited by the pace of photosynthesis and the amount of biomass. And so it took off, along with the human population.

Unfortunately, everything has side effects; and this story of human progress is no exception. With low-cost mass production after WWII, high technology has democratized. But it has also individualized. Personal laptop, smart phone, iPad, cars, and flat-screen TVs… In industrial societies, it is not unusual to see families where every member has a personal version of each. And new versions come out all the time, so having the old versions was not cool anymore, and you had to buy the newest ones.

What has grown alongside technology is a culture of mass-consumption for individuals. When companies invent and develop new gadgets, their dream is not that every community would have one. Their dream is that every family, every person would have one.

Growth in consumption and GDP has become *the* goal, and Brazil, China, India, Russia and almost everyone else have joined the quest. Young people everywhere now grow up with the dream of one day living like an well-to-do American… Having their own houses, their own sports cars, many televisions, and the latest electronics. And the dream is coming true for more people than ever before.

But unfortunately, and I think most of us here know it, this particular story of human progress is unlikely to have a happy ending. More and more individual consumption in a world of 7.3 billion people–going to 9.7 billion by 2050–will be a disaster.

Even the basic math doesn't work. The Global Footprint Network estimates that we are consuming 40% more resources than the Earth can produce–and 440% more if all of us lived like Americans. We are already losing species at about 1,000 times the natural rate. In 2011, the WMO declared 2010 the warmest year on record, but the record was broken again and again and last year 2016 was the warmest year ever, and recent scientific studies by many institutions suggest the world may get 4–5 degrees warmer in this century, if we keep business as usual. Extreme weather is getting stronger and happening more often everywhere.

What is scary is that all of this is happening with today's human ecological footprint. To believe that everything will be fine would be the most naïve wishful thinking. Nature is already warning us.

We can only come to one conclusion: the model of development started by the rich countries and adopted by the developing nations is not the right one. Individual consumerism is not the best use of technology. If we wish to remain on this Earth for centuries to come, then we have to find other ways to develop and improve people's lives. The rich countries are already "over-developed." The developing countries, or "not-yet-over-developed" countries, cannot follow them. They have to find better ways.

I firmly believe that we can find ways of development that are both better and more sustainable. I say "ways," not "way", for many reasons. For one, every place is different, and calls for development that's appropriate to it. An Indian way of development would be different from a Moroccan way of development; and how you develop in Crete would be likely be a little different from how you develop in Taiwan. It is extremely important for us to accede to the fact that human society has been over loading the earth system over the last half a century and we need to reexamine what "Development" really mean.

The difficult problem of climate change we are facing today is a global problem and so it needs a global solution. Neither a single country nor scientists can solve this problem alone. The earth does not care which community in which country puts out how much CO_2. It cares about the total amount of greenhouse gases, from all humanity. In December 2015, 195 political leaders from all over the world came to Paris to attend the COP21. The final agreement to limit the global temperature rise to $1.5–2.0°C$ is a historical great awakening. For accomplishing this goal, they also agreed that it is necessary for human society to decarbonize and to become carbon neutral in the second half of this century. Although agreement reached in COP21 is not a binding one, every Country is supposed to follow their committed plans.

Obviously, we will also need better global institutions. Nation state based international organizations are often mired with the dilemma between "collaboration" and "competition". Scientists believe that if we instill sufficient funding for research related to energy transformation, storage and transportation, it is likely that we can accomplish carbon neutrality by the middle of this century. By 2050, it is likely that the worsening global warming will bring frequent serious extreme weather events, that humanity would finally realize that the real danger of human survival is not the invasion of enemies across their national boarder. It is the climate change which we are causing it ourselves. Apparently, the 2% of global GDP which we have been spending on defense worldwide should be spent on research for the energy transformation. We should learn to defend ourselves from the "real enemy".

I do believe that for the global sustainability, we have to follow the following pathways,

(A) Global response to global problem.
(B) Back to nature, back to sunshine.
(C) Live better for less.
(D) Control population explosion.
(E) Improve equality around the world.

We do not have much time left. Unless we learn to connect humanity as one global unit, connect knowledge to action immediately, survival and prosperity of humanity will be extremely difficult.

Some Security Issues of Living on Planet Earth

Yannis A. Phillis

Civilization is a very recent event in Earth's existence. It became a reality, among others, through a combination of Darwinian individual and group competition. Narratives of both types of competition can be found in myths and stories in various sources including the Bible. If the time span from the appearance of the Earth 4.6 billion years ago to the present is equivalent to one year, then the Genus *Homo* which is 2 million years old appeared at 20:00 h, December 31. *Homo sapiens* (60,000 yrs ago), appeared 7 min before midnight and civilization (10,000 yrs ago), just 69 s before midnight.

This recent event of civilization afforded us enormous progress in such areas as life expectancy, economic prosperity, arts, science, and civic comfort, to name but a few. Today civilization has become global in that humanity is interlinked in complex ways. Economic troubles on national scales tend to have global repercussions, international trade has the potential of benefitting and, sometimes, harming everyone, greenhouse gases (GHG) emissions in one place affect climate everywhere, viruses and bacteria travel fast from country to country.

Civilization itself is like a thin skin, which, when torn, reveals the primeval human animal. Take away basic food and water security or infuse a sense of national or tribal encirclement by enemies or convince people of their national or religious superiority and you've created out of educated people with fancy degrees monsters and cannibals—the latter sometimes literally as was witnessed in, for example, WWII and SE Asia in the 1940s. ISIS is a case in point presently but not the only one.

Civilization and progress rely on the economy that, in turn, relies heavily on environmental services. But the economy creates externalities such as climate change, species extinction, disruption of natural cycles, large scale pollution and the list goes on.

Y. A. Phillis (✉)
Technical University of Crete, Chania, Greece
e-mail: phillis@dpem.tuc.gr

© Springer International Publishing AG, part of Springer Nature 2019
J. A. S. Kelso (ed.), *Learning To Live Together: Promoting Social Harmony*,
https://doi.org/10.1007/978-3-319-90659-1_2

The economy is central in creating utility and progress. Its main vehicle is what is called the free market. Absolutely free markets and Pareto optimal equilibria exist only in idealized economics courses and fantasies of the ultra rich. An absolutely free market would wreak havoc on the environment and the livelihoods of billions of humans.

It becomes clear, therefore, that to live more or less harmoniously on planet Earth, we need to rethink several aspects of our civilization and define a rather sustainable course. My modest proposal touches upon one major issue, climate change, which, of course, is only part of a long list of possibilities.

We urgently need to take immediate and drastic GHG abatement measures to prevent possible irreversible catastrophes in the near future. We are already experiencing more frequent droughts and extreme weather phenomena and the reappearance of long forgotten diseases. I shall analyze specific policies for all the countries based on my group's research.

Civilization according to the Dictionary of the Academy of Athens is "The body of moral, intellectual, material and technical achievements of a society." It is characterized mainly by: writing, urbanization, the existence of social hierarchies and the domination over and separation from nature. Civilizations rise and fall. Some reasons of collapse according to Jared Diamond (2011) are: environmental catastrophes (cutting of forests, soil erosion, etc.); climate change; resources dependence on remote places; internal or external conflict; no reaction to social and environmental problems.

In the past several advanced civilizations collapsed because humans destroyed their social and environmental base. Well-known examples of great civilizations that collapsed in the past include the Minoan, the Mycenaean, the Sumerian, the Babylonian, the Roman, the Mayan and numerous other civilizations (Motesharrei et al. 2014). One might wonder if we have not taken a similar path today because of climate change.

The Climate Predicament

Climate change is one of the most urgent problems facing the earth. Its facets are multiple: environmental, economic, and social, and its consequences could become dire if drastic and concerted action is not taken immediately. Climate change is already exerting a host of stresses on the environment and the society that will intensify with time. In the face of this reality, humanity has done little to avert possible catastrophes. Puzzling as this behavior might appear at first, it can be partially explained by behavioral economics and psychology (Grigoroudis et al. 2016). However, human climate inaction is often attributed to a degree to plain economics. We stand to lose economically more than we gain. Is this true?

Models that estimate the costs of climate change have been amply criticized on various grounds. An oft cited criticism is that they underestimate the costs of climate change. Ironically they show that this cost change could become astronomical reaching about 5% of global GDP by the end of the century (Kanellos 2014).

An interesting fact about climate change and the economy is that there is a negative correlation between income and climate concern. The higher the share of a nation's global CO_2 emissions the less its citizens and governments are willing to reduce them.

One could posit that ignorance about the facts of climate change is one of the main causes of global inaction. However, as levels of information about climate change increase, the willingness to take personal responsibility and action decreases.

People often realizing the enormous scale and complexity of the problem prefer to distance themselves from it. Redeployment of attention plays an important role. People push back the unpleasant facts about climate change and redeploy their attention to other routine things. People and, by extension, governments sometimes resort to a number of strategies to justify inaction. They isolate themselves from unpleasant information or blame others of doing a far greater damage or turn reality upside down and present polluting policies as environmentally benign. For example Norwegians justify oil exploitation on grounds of Norway being "a little land" whereas the US is causing a lot more climate harm (Hovden and Lindseth 2002). Going one step farther, Norway is supposed to be doing the environment a service because Norwegian oil is less polluting per unit volume when compared to oil from other places. More importantly, ignoring scientific facts and showing no tolerance for scientific uncertainty, no matter how small, lead to a kind of false collective rationality that justifies burning fossil fuels.

Some barriers that prevent us from acting are:

- ancient brain that has evolved to deal with immediate dangers
- when humans hear a lot about the problem they tend to become numb and insensitive
- scale: most people believe that their impact will be minimal given the global scale of climate change
- ideologies, political (conservative, liberal) or religious (God will not permit total destruction),
- reactance by people who mistrust science
- or have a strong interest in the fossil fuel industry.
- one could view climate inaction partially as the result of the bystander effect.

The UN has set the goal of containing temperature rise to 2°C above preindustrial times by the end of 2100. The question arises if this goal is attainable in some optimal sense and what reductions of GHGs are needed to achieve it. The answer to this is found in (Grigoroudis et al. 2016) and outlined below without mathematical technicalities.

If emissions of fossil carbon are abated so that a given measure of life satisfaction is optimized, then the trajectories of Fig. 1 are obtained for 16 regions that cover the whole earth.

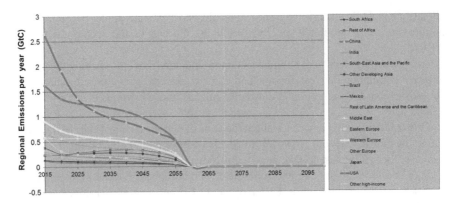

Fig. 1 Regional emissions cuts

Figure 2 shows the per capita carbon emissions and Fig. 3 shows the corresponding temperature rise for the period 2015–2100. Optimal emissions reductions are quite sharp in the period 2015–2060, with an average abatement of 1.05 GtC per 5 years. In this optimal scenario the 2°C becomes elusive since temperature rises to 2.5°C. Under this optimal scenario, climate damages rise from 0.48% of the global GDP in 2015 to 4.08% in 2100. No optimal scenario leads to 1.5 or even 2°C, which are the ambitious goals of the UN. Since no sharp reductions are expected before 2020 or even later, the 2-degree target becomes even more remote.

The largest overall emissions reductions should be undertaken by China, the world's biggest polluter, followed by the USA, one of the biggest polluters per capita. China's emissions should decrease from 2.66 GtC in 2015 to about 2 GtC in 2020, while the USA's and Western Europe's emissions should decrease from 1.63 and 0.91 GtC in 2015 to 1.38 and 0.73 GtC in 2020, respectively. Regional per capita emissions have similar trends (see Fig. 2). The annual per capita emissions in

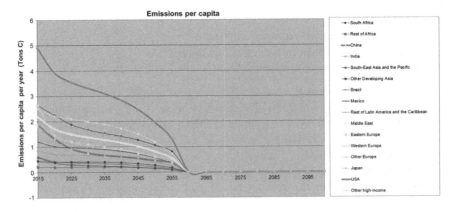

Fig. 2 Optimal emissions reductions per capita

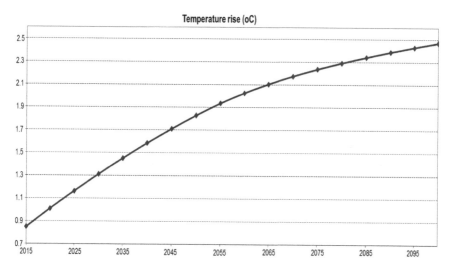

Fig. 3 Temperature rise under optimal scenario

the USA must decrease from 4.91 tons C per capita in 2015 to 4 tons in 2020. In other regions with high per capita emissions, such as Eastern Europe, South Africa, Japan, and the Middle East, reductions vary from 2.5 tons C per capita in 2015 to 1.5 up to 2.3 tons in 2020.

The China-USA Scenario

In December 2015 all nations of the earth signed the Paris Climate Agreement, with the exception of Syria and Nicaragua. In this Agreement nations made the so-called Intended Nationally Determined Contributions (INDCs) to carbon reductions. The USA made quite concrete proposals whereas China's commitment is rather fuzzy. However, China is taking action towards reducing its carbon footprint while the USA pulled out of the Agreement. In what follows, the model of (Kanellos 2014) was run assuming that both INDCs were valid to obtain an idea of the future. China's INDCs are as follows:

- Peaking of CO_2 emissions around 2030 and making best efforts to peak early.
- Lowering CO_2 intensity (CO_2 emissions per unit of GDP) by 60–65% from the 2005 level.
- Increasing the share of non-fossil fuels in primary energy consumption to around 20%.
- Increasing the forest stock volume by around 4.5 billion m^3 from the 2005 level.

Combining China's emissions pathway in 2015–2100 with that of the USA and incorporating both into the optimization model, while all other regions are

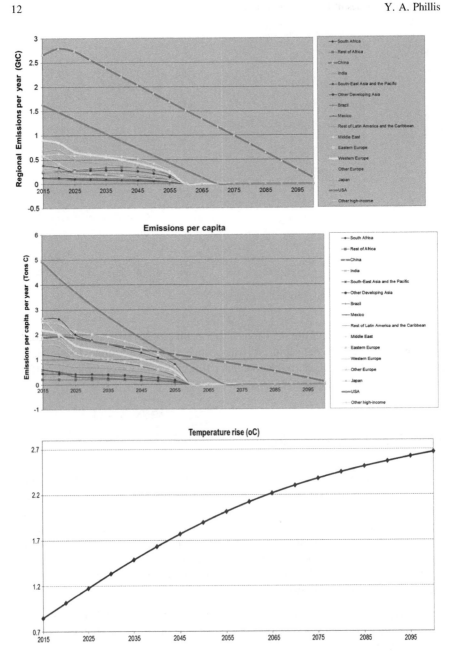

Fig. 4 Total and per capita carbon reductions and resulting temperature rise

optimized, the results of Fig. 4 are obtained. This scenario leads to a reduction of global emissions from 9.46 GtC in 2015 to 8.37 GtC in 2020 and 1.97 GtC in 2060. The temperature rise in 2100 will be $T_{2100} = 2.65°C$ and it will continue to increase at a rate $T_{2100} - T_{2095} = 0.05°C$. The global damages are larger in this scenario reaching 4.35% of world GDP in 2100. Once more the elusiveness of the two-degree goal is highlighted.

Optimal Results with Negative Emissions

The Paris Agreement relies, among others, on negative emissions, that is technologies and policies that sequester carbon, such as CO_2 capture and storage or land afforestation. Introduction of negative emissions changes the previous curves a little as seen in Fig. 5. Temperature rise now is estimated at 2.47°C in 2100 and exhibits a rather stable pattern about this value because now $T_{2100} - T_{2095} = 0.02 < 0.025$. Additionally, the initial mitigation time is postponed for the year 2020 because negative emissions permit a less drastic abatement policy. Global optimal emissions increase from 9.45 GtC per year in 2015 to 10.53 GtC in 2020. Then substantial mitigation action should be undertaken in order to reduce global emissions in 2060 to zero. Emissions between 2020 and 2025 should be reduced by 27.5%. Past 2060 global emissions become negative all the way to 2100, reaching—1.77 GtC per year at the end of the period. Finally, the estimated global climate damages are similar to the optimal scenario with positive emissions (0.48–4.2% of global GDP).

At first glance, the negative emissions scenario appears promising but in reality it is based on future technological advances that might prove to be false. Several authors question the premise that large-scale negative emissions are technically, economically, and socially viable. The most important current negative emissions technologies include:

(a) Reforestation and afforestation (conversion of land into forest): They are not technologies in the strict sense, but they are important mitigation strategies. Although trees can draw an amount of carbon from the atmosphere, the process is slow and it requires large land areas in order to make a significant difference to global CO_2.

(b) Bioenergy with carbon capture and storage (BECCS): Currently, it is the most widely known negative emissions technology. Presently no large-scale BECCS can be deployed, while their potential is not limitless.

Other negative emissions technologies are currently in different development stages, but their cost and energy intensity would likely be high (e.g., direct air capture process).

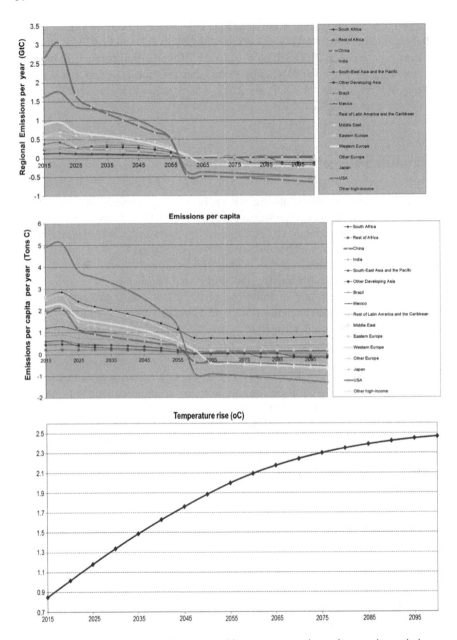

Fig. 5 Optimal emissions trajectories and resulting temperature rise under negative emissions

From the exposition above it can be seen that:

- the UN 2°C target that would probably confine global warming to tolerable levels is elusive
- very drastic measures must be taken immediately anyway to avoid great catastrophes
- public cognition of and attitude towards climate change should drastically change

Some Proposals

Given the brief analysis above I propose the following:

- The Paris Agreement, a good first step as it might be, needs tightening fast
- hydrocarbons, natural gas, and coal must be abandoned as soon as possible in favor of renewables
- It's high time for the universal introduction of electric cars
- proper energy pricing/taxation
- provide citizens with incentives to insulate/aerate homes
- Education of the public so that people:

reduce meat consumption
use public transportation extensively
reduce superficial consumption
press politicians to take climate action

We need governments that are strong, enlightened, and independent of special interests to achieve the goals above. In today's complex and precarious world we have not the luxury of anything less. "Politics" today is a profession, often devoid of the true meaning of politics, which is serious debate and deliberation about social problems. It has primarily been reduced to a struggle for political power and means. This must change when the very foundations of our civilization are at stake. We know enough about climate change and how to control it technologically. We need political will to act immediately. Time has run out. If we wait another decade discussing the problem, we should expect temperatures of 3°C above preindustrial time with dire consequences. Action is summarized in four words: **abandonment of fossil fuels**. This will certainly happen sooner or later. The question is if it will happen before our fall into the precipice. The problem is urgent. It remains to be seen if our leaders will see it as such or we'll go down in history as those who destroyed their children's future.

References

Diamond, J. (2011). *Collapse: How Societies Choose to Fail or Succeed*. New York: Penguin.

Grigoroudis, E., Kanellos, F. D., Kouikoglou, V. S., & Phillis, Y. A. (2016). Optimal abatement policies and related behavioral aspects of climate change. *Environmental Development, 19*, 10–22.

E. Grigoroudis., F. D. Kanellos., V. S. Kouikoglou., & Y. A. Phillis. The challenge of the Paris Agreement to contain climate change. Intelligent Automation and Soft Computing, to appear.

Hovden, G., & Lindseth, G. (2002). Norwegian climate policy 1989–2002. In W. M. Lafferty., M. Nordskog., H. A. Aakre (Eds.), Realizing rio in norway: evaluative studies of sustainable development. Oslo, ProSus, pp. 143–168. http://www.prosus.org/publikasjoner/Boeker/ Realizing_Rio.htm. Accessed 19 April 2016.

Motesharrei, S., Rivas, J., & Kalnay, E. (2014). Human and nature dynamics (HANDY): Modeling inequality and use of resources in the collapse or sustainability of societies. *Ecological Economics, 101*, 90–102. (May)

Kanellos, F. D., Grigoroudis, E., Hope, C., Kouikoglou, V. S., & Phillis, Y. A. (2014) Optimal GHG emissions abatement and aggregate economic damages of global warming. *IEEE Systems Journal*, to appear.

Mind Wars: Brain Research and National Defense

Jonathan D. Moreno

Why would one think of neuroscience as a fertile field of medical ethics and national security? That is the question J answer in my recent book, Mind Wars (Dana 2006).

Neuroscience has almost surely grown faster than any other interdisciplinary area in the past decade. The Society for Neuroscience is host of one of the biggest science meetings in the world, drawing about 40,000 attendees from disciplines including neurology, psychology, computer science, radiology and psychiatry, as well as my own field of bioethics.

My fascination with the ethics of neuroscience research is rooted in two distinct experiences. First, I wrote a book called Undue Risk: Secret State Experiments on Humans (2001), about the history and ethics of human experiments conducted for national security purposes. My work on that book-in addition to my role as a staff member for a presidential advisory commission on radiation experiments on humans-sensitized me to the complex relationship between science, ethics and national defense needs. Understanding that science is essential for US military superiority, every presidential administration since World War II has provided federal largesse to support research, creating a relationship between academe and government that has served the country well.

Second, as I followed developments in neuroscience through the popular press and scientific publications, I noticed that many of the most exciting experiments were supported by national security agencies, such as the Defense Advanced Research Projects Agency. Yet press coverage of neuroscience experiments usually mentions the source of funds only in passing.

This paper is adapted from Moreno, J.D. "Brain Research and National Defense," The Chronicle of Higher Education, November 10, 2006.

J. D. Moreno (✉)
David and Lyn Silfen University Professor and Professor of Medical Ethics and of the History and Sociology of Science, University of Pennsylvania, Philadelphia, PA, USA
e-mail: morenojd@mail.med.upenn.edu

© Springer International Publishing AG, part of Springer Nature 2019
J. A. S. Kelso (ed.), *Learning To Live Together: Promoting Social Harmony*,
https://doi.org/10.1007/978-3-319-90659-1_3

I wondered what security agencies' financial support might say about their interest in the long-term contributions of neuroscience to national security. Though that question seemed to me to be the 800-pound gorilla in the neuroscience lab, to my amazement, no one else, including the neuroscientists, appeared to be asking it in any systematic way.

In one important sense, that is perfectly understandable. Scientists focus on their particular research aims, not on the long-range interests of financial supporters. What seems to an investigator to be a very limited research question can be seen by a security agency as part of a larger pattern. And, of course, even scientific advances that do not stem from military-sponsored research can be adapted for security purposes later.

Not all the neuroscientists I spoke to were enthusiastic about discussing such issues on the record, to put it mildly. Their reluctance only served to confirm my sense that those matters were of more than passing interest, and led to my new book, Mind Wars: Brain Research and National Defense.

Fortunately for my project, not all scientists or former agency employees were unwilling to talk to me, and much information is a matter of public record. I came to appreciate the way that DARPA, in particular, does business, for it is a science agency, not a spy agency, and the vast majority of its work is done in concert with scholars. Thus, although many of the studies that raise interesting ethical and social questions are sponsored by DARPA, that does not imply that the agency should not be supporting them, or that the research should not be done.

Of course scientists in any field are understandably reluctant to make comments that could jeopardize future financial support for their work, but there is a special sensitivity to being harnessed by the US military for national security, with alarming ethical implications-in some cases, almost paranoia-about the suggestion that scientific research is leading to mind reading or mind control. That sensitivity is partly left over from the early days of the Cold War, when US government officials suspected that treasonous statements by prisoners of war in North Korea were the result of brainwashing. Twenty years later, we learned that the CIA and the US army had themselves engaged in mental manipulation experiments that included the use of hallucinogens such as LSD. People read each other's minds all the time, sometimes unconsciously and relying on various cues such as body language, and our minds are controlled in countless ways, from natural stimuli such as odors to pop-up web ads. But most of us get nervous when we imagine that some distant authority could have access to what we like to think of as private thoughts, or that some deliberate and fairly precise means can be used to alter our cognition or behavior in accord with someone else's strategic purpose.

I don't know if any of the contemporary research projects that I discuss in my new book qualify as mind reading or mind control, but some of them seem pretty close. Certain brain-scanning techniques, especially functional magnetic resonance imaging, have stimulated a huge amount of research attempting to correlate neural activity with specific tasks or experiences.

In one famous and contentious study, negative automatic responses by white research subjects to photographs of black faces were correlated with activity in the amygdala, which processes emotion in the presence of stimuli. Or take the example of "prisoner's dilemma" experiments, in which both subjects benefit when they co-operate with each other. When the subjects do well, neurotransmitters activate pleasure centers in the brain.

Some neuroscientists claim that fMRI can already show when subjects are thinking of a certain number, when they are lying, or what their sexual orientation is, and that the technique will make even more refined and precise analyses possible in the years ahead.

Other studies focus on replacing old-fashioned lie detectors with systems based on neuroscience. The hope is that the new techniques would not only be more reliable, but that they could replace torture and other physically aggressive means of interrogating terror suspects and enemy operatives.

It's not hard to see additional security implications of such technical capabilities. For instance, a spy agency could measure the neurotransmitter secretions of candidates for special missions, to see how they react to stress. Military personnel in information-rich environments, such as cockpits, could have their brain functions monitored for information overload, and officers behind the front lines could modify the flow of data accordingly, using devices now being developed to provide real-time remote brain imaging.

Other direct interventions to enhance soldiers' capabilities could come in many forms, including new generations of neuropharmaceuticals, implants, and neural stimulations. New anti-sleep agents such as modafinil (which, under the brand name Provigil, some students may already have discovered) are replacing old-fashioned amphetamines among fighter pilots as well as globe-trotting business executives. DARPA's "peak soldier performance" program aims to improve metabolism on demand so a soldier could operate at a high level for three to five days without needing sleep or calories, except perhaps high-nutrition pills.

DARPA is also interested in increasing the "bandwidth" of soldiers' brains. One idea is to develop something called a brain prosthesis, a chip that–if it could be made to work–would restore mental functioning in people who have epilepsy or have had strokes. But experts disagree about whether such a device, intended to treat a medical condition, could also improve normal mental functioning.

Or perhaps extra copies of genes that code for certain neural receptor sites could be introduced into the brain to improve learning skills; that has been done in mice, in the lab of Joe Z. Tsien of Boston University. Electrical stimulation has been used with some success as an adjunct to standard rehabilitation techniques for stroke victims; could it improve cognitive functions in healthy individuals?

Intelligence and endurance are not the only traits that make a good soldier. Another is the ability to manage fear. In an interesting experiment, Gleb Shumyatsky's research team at Rutgers University in New Jersey, found that mice bred not to have the gene stathmin did not exhibit normal fear behavior, such as freezing in place, as often as normal mice did when exposed to things such as a mild shock.

Stathmin is expressed in the amygdala and is associated with innate and learned fear. The mice without stathmin froze less often because they had impaired learning capacity.

It is unlikely a particular gene in humans corresponds so precisely to fear, given differences in the way mouse and human genes are expressed. But given past hype about, say, the so-called gay gene, it is easy to imagine an overly enthusiastic official proposing to screen recruits for the "fear gene".

Where there is fear, there is often long-term trauma. Trauma victims who had been given the beta blocker propranolol-which is normally used to treat heart disease but which inhibits the release of brain chemicals that consolidate long-term memories with emotion-scored lower on a scale measuring post-traumatic stress disorder than did members of a control group after a month of psychological counselling. The difference was not statistically significant, but another result is more important. Three months later, none of the beta blocker recipients had elevated physiological responses when asked to recall their traumatic experiences, while 40% of the control-group members did.

Those results give hope to sufferers of post-traumatic stress disorder. They also raise the question of whether the drug could be given prophylactically, before a person enters what could be a traumatizing situation. How would we feel about preventing the disorder in young soldiers going into battle: preventing a lifetime of harrowing memories as well as the soldiers' capacity to connect horrific experiences with negative emotions? Do we really want guilt-free soldiers?

The hair on the back of readers' necks may be rising at the prospects ahead of us and their ethical and legal implications. How are we to ensure that interventions such as those to manage guilt and fear would be confined to military operations against truly dangerous adversaries and not more widely adopted, perhaps by civil authorities or criminals?

The importance and difficulty of regulation become even more impressive when we consider that many of the technologies would be not only useful tools in military situations but also promising advances in health care. The same sort of device that would allow an officer to see if a pilot was receiving too much information, say, could permit a nurse in a doctor's office to check up on the welfare of a brain-injured patient at home. There are also commercial possibilities, of course. For instance, businesses are already intrigued by the possibilities of using brain-imaging techniques to conduct market research.

Much of the history of bioethics might be read as a 40-year conversation about the prospects for changing human nature through startling developments in the life sciences. Bioethicists have largely played down such concerns, noting the extent to which we already deliberately change ourselves in all sorts of low-tech ways, such as using sleep medication or taking French lessons.

However, an alternative view has recently been getting more attention. Its supporters-including Leon R. Kass, professor in the Committee on Social Thought at the University of Chicago and former chairman of the President's Council on Bioethics-contend that practices such as new reproductive technologies, while attractive and seemingly benign, have profound but unpredictable societal implications. The debate about what types of enhancement are permissible, given the risks, should be expanded to include the sorts of interventions I have described, especially given the clout of national-security funds and goals.

The defense implications of neuroscience also raise policy questions about civil liberties, regulation and safety. We're familiar with the role of atomic scientists in the control of nuclear weapons, and more recently biologists have become key players in planning for defense against bioterrorism. The same is not yet true of neuroscientists, partly because the idea that neuroscience could be involved in national security is only now becoming clear, and partly because neuroscience is a complex interdisciplinary field whose practitioners work in separate silos. But the day is approaching when we will have to consider those issues in a more systematic way.

Programs in neuroscience are springing up at colleges and universities across the US. The programs should include discussions of science policy, such as: How do the sources of research funds affect the direction of science and social change? Which uses of brain research are acceptable and which are not? And what limits should society, perhaps acting through scientific associations, place on the acceptable applications of neuroscience?

Whatever the future holds for neuroscience, it would be naive to suppose that national-security organizations are not monitoring developments in that field as they do in any other. We need a public conversation about the role of brain research in defense.

Learning to Live Together: Linking European and Local Initiatives

Leonidas Makris

It is the very same nature of the current problems of the world that are asking for more unity between people. It is the context of our life and reality which is demanding a far more inclusive polity. In this direction, I will be suggesting ways on how to promote people's coexistence at two different levels: On the one hand, how to consolidate and make more viable in the future the European project, a potentially useful paradigm of coexistence for other continents and humanity. On the other hand, how to serve in practice the multiple needs people of very different backgrounds face where they reside, which for an increasing majority of them will be metropolitan urban areas. In addition, I will also suggest how collaboration at these two different levels could be coordinated and inten-sified in order to accommodate better our needs. If policies at these two different levels are attuned with the goal of bringing people closer together, we can hope to create a more inclusive and viable society. An ambitious precondition for this is to shift gradually the centre of political decision from the level of the nation-state to a more global and inclusive polity which, at the same time, is agile enough to assist and facilitate people at a local-city level. Taking into account the looming conundrums of globalisation, it is a matter of necessity, and survival I dare to say, to create a new two-folded system of governance which could meet the challenge and transform the world into a friendlier and better place.

There are many ideas and concepts which were invented in order to facilitate people's coexistence. A great number of them were exploiting the psychological need people have to belong somewhere: into a group, a clan, a nation, an empire (Tajfel and Turner 1986). However, the viability of the so called 'in-group'—which is usually opposed to an 'out-group'—depends on the needs that it serves and the problems it solves (Westin et al 2010). Our society during the last decades, due to

---

The author holds a PhD in Political Science from the London School of Economics. He is advisor to the Mayor of Thessaloniki and teaches at the American College of Thessaloniki

---

L. Makris (✉)
American College of Thessaloniki, Thessaloniki, Greece
e-mail: lmakris2@yahoo.com

© Springer International Publishing AG, part of Springer Nature 2019
J. A. S. Kelso (ed.), *Learning To Live Together: Promoting Social Harmony*,
https://doi.org/10.1007/978-3-319-90659-1_4

the unprecedented technological advancement of digital technology and communication, is currently in probably the most intensified phase of globalisation. This trend seems to be irrevocable and has very concrete repercussions. The problems that humanity is faced with are of a global scale and nature: Concentration of economic power and exacerbated inequality, more frequent cyclical and contagious economic recessions, overpopulation and scarcity of resources, environmental pollution and global warming, massive waves of immigration. It is all the more obvious that these issues cannot be tackled by individual national states. Strangely enough, and despite the conspicuous need to confront them in a different, more unified and coordinated manner, the reaction to the above mentioned threats has been fragmented at best (Lundestad 2004).

It is the very same nature of the current needs of the world that are asking for more unity between people. It is the context of our life and reality which is demanding a far more inclusive and, at the same time, familiar to the people polity. And it is a matter of necessity, and survival I dare to say, to meet the challenge and transform the world into a friendlier and better place. In order to advance this vision, I will be suggesting that it is time to simultaneously interrelate and strengthen different levels of governance: A supra-national paradigm which permits more efficient coordination and a spirit of unity, and a local level which is closer to the people's everyday reality and can revitalise their interest in participation and democracy.

Only in the aftermath of the worst disaster–WWII–some progressive Europeans came up with the idea of a united Europe. It was a great idea and despite the shortcomings and what many critics say, due to the creation of the Union its members are already enjoying the longest peaceful period in our continent. This is probably its most valuable achievement (Chatam House-Kantar Survey 2017). However, if we would like it to last, we need to work harder and invent ways of how to make the whole European project viable.

The aim of the present text is much more modest than that. But if we really want to learn how to live together, the European paradigm needs to succeed and become an example for other similar projects which have been springing up during the last decades in other continents. Thus, if Europe wants its union it needs to 'create' a European identity. A skeptic could reply that since national identities are already present and consolidated such a creation would comprise a distortion of history, a refutation of reality.

What we need to understand though is that even theorists who underline the importance of nations and the significance of ethno-symbolism (Smith 2009) claim that very often the creation of national identities is a kind of an 'ideological' construction. It is very often based on historically flawed interpretations of the past and is built on common myths which intend to initiate or intensify the bond of solidarity between the in-group members who share them. In other words, it might not be based on the most accurate scientific interpretations but the purpose of its creation aims to unify people (Ibid. 2009). And unifying people at a European level does not necessarily mean -as it is the case with national myths- that their identity should be construed in contrast to outsiders. Under certain circumstances this process can serve a benign goal.

It is therefore evident that if the EU would seriously decide to build a common European identity, this could be feasible and productive providing its members manage to isolate useful elements of our culture and history which could comprise our mutual psyche. We would need to decide what we really have in common, which are the cardinal ingredients that could make us feel proud of being Europeans and generate emotional ties between us. It is of the outmost importance to complement the 'sober' cognitive process which could lead to the decision of the elements comprising the core of our civilization with a more colorful myth which would allow powerful emotional attachments. We need to cultivate both instrumental as well as sentimental ties to our new European identity in order to strengthen our attachment to it (Breakwell and Lyons 1996).

This is certainly not an easy project, since it would presuppose to progressively manage to loosen our ties with our existing national identities. What we need for a trade off involving a gradual weakening of national identities, in return for gaining a stronger European one, is strong political will. For the sake of a better and peaceful future where institutions could accommodate better our needs and solve our problems, in national governments the political trend should counterintuitively be to reduce their sovereignty. EU institutions, according to the same unifying plan, should be willing to democratize and represent directly and equally people from all parts of Europe. They should, thus, move away from a compromising policy which primarily reflects the short-term interests of the most powerful EU members (Krotz and Schild 2013). As far as the long-term interests of every member are concerned, they would be better served if Europe will start acting as a real union serving in a more balanced way the interests of every one of its regions.

However, we have to be realistic. Right now the necessary political will in order to move towards this direction does not seem to be present at all. Most of the national governments attribute to the EU whatever seems to be unpopular among the electorate. In addition, in many cases they advertise as their own policy whatever EU policy is received well by the people. The crucial shift in the political will can only come gradually and from a bottom-up process. A new educational approach disseminated to all students of different levels and schools could help to contrive the new European identity. Stressing the historical periods and elements which brought closer the people of this continent could facilitate the forging of such identity (Van der Leeuw-Roord 2007). Individuals from an early age can learn and be proud of our common European roots originating from the classical period of history and culture, as well as from its refined expressions rediscovered during Renaissance, the age of Enlightenment and Modernity in general (Berger 2012). Despite the fact that the writing of any such common history aiming to shape a collective myth would presuppose gross oversimplifications, in this case, the process could serve a commendable goal. It is of the outmost importance to find ways to transcend our national myths and 'Europeanise' our historical consciousness. At least if we want to live in a better world. "Decentering our still predominant national histories and weakening the link between national histories and national identities remains very much a task for today if we want to build a less conflictual and less painful future..." (ibid. 2012, p.45).

Other aspects of the badly needed new European identity could always underline the humane way of organizing society, the creation and maintenance of the social state and the organic solidarity which connects social classes in our continent and transcends national borders (Bauman 2004). It goes without saying that a new and roughly common for all continental schools history should develop gradually. It should aim at converging national versions of history by accentuating the common European projects and alliances after abstracting them from the instances of national animosity between Europeans. We should not insist -as we now do- in emphasizing past ethnic conflicts manifesting the heroic feats of each national group to the detriment of the opponent, as if we were proud of our barbaric, at times, past (Berger 2012).

Taking into account that mutual efforts of two or more European countries to rewrite their history in more consensual ways have failed, the abovementioned project remains highly ambitious. Yet, I would dare to say that despite it being seemingly unduly ambitious, it is feasible. And it is feasible simply because it is the only way to move forward. Our lives are gradually more and more inter-dependent and our problems more and more mingled. For solutions to be effective they need to be common. See for example the challenging issues our monetary union has been facing and the difficult, albeit common, efforts to address them. In order to consensually move forward we need to build gradually our unity on solid grounds. And devising a common new European identity -of at least equal importance with the national ones- is a precondition to resolve our problems much easier than now and without blaming each other. It is a precondition in order to learn how to tolerate each other, strengthen our mutual understanding and, in addition, discover how to act in unity.

Notwithstanding the above, in Europe there are already many examples of everyday tolerance and consolidated understanding of our differences, as well as of peaceful and harmonious coexistence. Despite some deviant instances that have been observed during the last years, such a model of a colorful, tolerant and understanding society exists at a lower level. European cities, particularly metropolitan urban areas, have already been practicing such a paradigm for decades now. Both institutionally as well as socially. They have made serious and concerted efforts to facilitate all of their residents' life, independently of their origins, beliefs, ethnic background or national identity. It is true that many of the ideas that sprung in Europe during the age of the enlightenment and have been developing throughout modernity have been implemented at a local level. Big European cities have advanced and became much more tolerant and receptive to an amalgam of new populations. They are also trying to accommodate their various needs and incorporate them into a new urban in-group which exists independently of the national one. The effort to integrate people and render them indiscriminate rights and opportunities based on their humane attribute, their local presence and contribution should be continued and intensified (Garton Ash 2016).

What I am suggesting is that this effort needs to be reinforced by creating new institutional ways through which the EU would trust the closest to the people authorities. The local ones. And I mean municipalities and city authorities, not regions as it currently does. If we want to democratize the way we are governed and if we want people to participate in politics more actively it is much easier to convince them to get involved in issues that affect them everyday. This is easier to attain where people live. According to UN data, more than half of them live already in urban areas, and very soon–in few decades (by 2050)–the vast majority of people around the world will be living in metropolitan areas (UN 2015). That means that cities should have more power to manage funds and EU help than rely on national or regional authorities. They should have more tools to assist their residents and convince them that it is worthwhile participating in the political as well as the actual management of the area of their dwelling.

Transforming the city to the locus and nucleus of European democracy would constitute a radical change. A beneficial evolution because, as I said in the beginning, institutions in order to be useful they should adjust to people's needs, the current and future ones, not the past ones. And while supra-national and local institutions can provide future solutions according to people's needs, national ones seem to me to be losing the meaning and the significance they had in the past (Barber 2013). While the transition to the new forms of governance should be gradual and not abrupt, insisting on schemes which serve past and not future needs could be regressive if not detrimental. You cannot extensively collaborate with someone if national stereotypes imply that he/she is inferior, less efficient or even base and worthless (ibid. 2013).

It is necessary to find institutional ways to create an extensive, viable and functional network of cities. Leveraging governing at a local level will not only rejuvenate democratic participation. Coordination between cities will create new ties and find instructive ways to bond people who can understand that they have to encounter similar problems. Cities around the world can create unions in order to manage common funds and projects and learn from each other. In particular, in the EU, city-level governance can become the front-desk of EU policy: manage social funds and create economic growth and jobs matching better the local needs while protecting the local environment. The initiative of the European Metropolitan Authorities (EMA) has created a forum to promote an agenda of common goals and challenges for big urban areas. It is certainly in the right direction but it needs to be transformed into a powerful institutional tool. A process attempting to link EU cohesion policy and its management to large urban places could aspire to comply with the ambitious EU economic, social and environmental goals by sticking to the principle of subsidiarity (Treaty on the EU 2012). This initiative could in the future become an alternative of a more immediate democracy which could make more personal and recognizable the currently bureaucratic and intricate scheme of regulating affairs. In addition, apart from offering new solutions to manage funds and projects, the new city-network could provide the locus within which more benign competition can advance.

People are by nature competitive and unfortunately sometimes aggressive. Yet, if we manage to channel their tendency to impose themselves on others to a harmless constructive process, this will release their tension. City-competition as we have seen in the past is by nature much more benign than national one (Barber

2013). Thus, it is necessary that cities should be represented much more robustly than what is now the case at a European level. Their saying should be determining things and not have an advisory role as it is currently the case.

The suggested combination of fortified supra and infra national schemes of government could become an example not only for Europe. Judging from how fast the world is changing and how pressing world-wide challenges are becoming, we are not far from the era when creating a global village–or a 'cosmopolis' as Garton Ash (2016) coined it- will not be only a necessity, it will be the only way to survive. Helping in order for the European project to succeed can make our continent a useful and peaceful paradigm for people to learn how to live together. And if we always need to create enemies in order to form and unify our in-group (Smith 1992), let these enemies be extra-terrestrials and not something more real and tangible.

References

Barber, B. R. (2013). *If mayors ruled the world: Dysfunctional nations, rising cities*. New Haven and London: Yale University Press.

Bauman, Z. (2004). *Europe: An unfinished adventure*. Cambridge: Polity Press.

Berger, S. (2012). Denationalizing history teaching and nationalizing it differently! Some reflections on how to defuse the negative potential of national(ist) history teaching. In M. Carretero, M. Asensio, & M. Rodríguez Moneo (Eds.), *History education and the construction of national identities*. Charlotte (NC): Information Age Publishing INC.

Breakwell, G. M., & Lyons, E. (Eds.). (1996). *Changing European identities: Social psychological analyses of social change*. Oxford: Butterworth-Heinemann.

Chatham House-Kantar Public survey. (2017). *Attitudes towards the EU-General Public*, Fieldwork, Fieldwork was carried out online between 12 December 2016 and 11 January 2017.

Consolidated versions of the Treaty on European Union. (2012)/C 326/01. http://eur-lex.europa.eu/legal-content/en/TXT/?uri=CELEX%3A12012M%2FTXT

Garton Ash, T. (2016). *Free speech: Ten principles for a connected world*. London: Atlantic Books.

Krotz, U., & Schild, J. (2013). *Shaping Europe. France, Germany and embedded bilateralism from the Elysee Treaty to twenty-first century politics*. Oxford: Oxford University Press.

Lundestad, G. (2004). Why does globalisation encourage fragmentation? In *International politics*, 2004 (Vol.41, pp. 265–76). Palgrave Macmillan Ltd.

Smith, A. (2009). *Ethno-symbolism and nationalism: A cultural approach*. New York (NY): Routledge.

Smith, A. (1992). National identity and the idea of European unity. In *International affairs*, (Vol. 68, Issue 1, pp. 55–76). (January).

Tajfel, H., & Turner, J. C. (1986). The social identity theory of intergroup behaviour. In S. Worchel, & W. G. Austin (Eds.), *Psychology of intergroup relations* (pp. 7–24). Chicago, IL: Nelson-Hall.

UN. (2015). *World urbanization prospects: The 2014 revision*. New York.

Van der Leeuw-Roord, J. (2007). A common textbook for Europe? Utopia or a Crucial Challenge? http://www.culturahistorica.es/joke/textbook_for_europe.pdf.

Westin, C., Bastos, J., & Dahinden, J. (Eds.). (2010). *Identity processes and dynamics in multi-ethnic Europe*. Amsterdam, Netherlands: Amsterdam University Press.

Part II
Diversity and Integration

The Human Group Instinct as Basis of Culture and Atrocities

Christoph von der Malsburg

On September 2, 1666, a fire broke out in a bakery in Pudding Lane in London, and within three days the City of London inside the old Roman City burned down. There had been fire prevention laws, but they had not been observed properly, there had been fire pumps, but they couldn't reach the fire in time, there had been plans of action, but they were not enacted forcefully. Today, there are strict fire prevention laws, smoke detectors, fire hydrants, professional fire fighters and emergency plans, and at least in modern states fire is no longer a major threat. Likewise, virus infections, earthquakes and climatic change are taken seriously, and great pains are taken to deal with these dangers. But there is another danger, greater than them all, that the world is turning a blind eye to. Virtually without interruption it is striking somewhere in the world in one place or other, and it is destroying lives and values at a grand scale. The names of Hitler, Stalin, Mao, Pol Pot, Srebrenica, Rwanda or Darfour all stand for gigantic disasters, but evidently the world refuses to take the message. These conflagrations are taken as isolated historical derailments "that cannot possibly happen here". There is painstaking research into the particular conditions which led to the Nazi Regime, but this only clouds the realization that it doesn't matter whether a fire is started by a match, a lighter or a lightning stroke and that the real issue are the mechanisms of the ensuing chain reaction. Group violence is made possible by the biology of human nature, which we all share.

The desert locust Schistocerca gregaria lives as a rather innocuous little insect until, under some specific conditions and high enough population density it changes its color, aspect and behavior and takes off in gigantic swarms to devastate large swathes of land. Similarly, humans, under appropriate conditions, are turned from innocuous, friendly citizens into rapacious beasts capable of just any sadistic atrocity imaginable. This behavior has been described by Le Bon (1952) and Freud

C. von der Malsburg (✉)
Frankfurt Institute for Advanced Studies, Frankfurt, Germany
e-mail: malsburg@fias.uni-frankfurt.de

C. von der Malsburg
University of Southern California, Los Angeles, CA, USA

© Springer International Publishing AG, part of Springer Nature 2019
J. A. S. Kelso (ed.), *Learning To Live Together: Promoting Social Harmony*,
https://doi.org/10.1007/978-3-319-90659-1_5

(1975), has been reflected upon by Gassett (1932) and Canetti (1973), has been described in very painful detail by Arendt (1968) and entertaining breadth by Bloom (1977), there are laboratory experiments (Milgram 1963; Banks and Zimbardo 1973) and classroom demonstrations (Jane Alliott's blue eye brown eye experiment) to show the basic effect, but all of this has not been able to arouse the collective attention that it deserves. There are international laws and an International Court of Justice, there are UN and non-government organizations to prevent or limit or mitigate the effect of group violence, but again and again all of this turns out to be too little too late, just as in the Great Fire of London. Like the environment, the climate or poverty, the group violence phenomenon will need a great worldwide rally and efforts to understand the phenomenon and to come up with effective measures to reign it in.

At the heart of the matter is our biological nature. I would like to talk of our *group instinct,* although scientists cannot agree yet on what it is (and cannot even agree on a definition of the word instinct). Our group instinct makes us do what is expected from us by the social group of which we feel we are part. Our group instinct is stronger than our instinct of self-preservation. Soldiers walk singing into war and death, and daily we read about suicide bombers. That our nature permits such behavior can only be interpreted in terms of altruism on a grand scale. We are ready to defend the existence or the superiority of our group with our life. Science is not sure how to explain the origin of this behavior. The obvious evolutionary mechanism to create it is group selection: our ancestral clan in the savanna made it against competitor groups by being fiercer in its aggression and self-defense, but science hasn't been able yet to come up with a concrete statistical model for how this might have happened. *Kin* selection, which can favor altruism toward others with whom we are likely to share many genes is an accepted mechanism, but the nation for which a soldier is ready to die is not a homogeneous genetic group. The best we can do at the present time is to accept the existence of the group instinct and try to understand its nature and its consequences for our communal lives.

The dominant aspect of the group instinct is that it induces us to do whatever our reference group expects from us. As such, the group instinct is instrumental in constituting organized society, civilization and culture. Without a group instinct we wouldn't be domesticated, we would lead solitary lives in the woods. Our urge to conform makes us to adopt the language, the style, the fashion and mores of our reference group, incites us to tremendous efforts to build our social standing and to maintain honor. In fact, our group instinct turns us into cells of a super-organism, man, which is the true author of economy, science, religion, sure knowledge and all the highest values we can think of. This is the inside, the benign face of the group instinct, the side that makes us live harmoniously together. We cannot possibly wish it away, even if some advanced genetic engineering made it possible. The flip side of the group instinct, the external face of it, is the potentially very ugly behavior that it incites in clashes between groups. There then are no limits to the cruelties of which we are capable. Hannah Arendt describes the regimes of Hitler and Stalin as grand-scale human experiments to find out or demonstrate that there are no limits whatsoever to the intensity and scale of atrocity of which humans are capable. Science is at the present time coming more and more to the conviction that our behavior is guided intensely by emotions and drives, that we don't come into this world as blank slates onto which rational mind is free to write rational behavior. It is rather our instincts, emotions, our *elan vital,* to give us direction and purpose. All rational thinking is but a servant to fit our emotions to the realities of modem life.

Are we, then, helpless victims of our biological nature, do we have to live with the negative side of the group instinct, as we cannot live without its positive sides? To whom can we turn for help? Science–in the form of neurosciences, social psychology etc. is certainly called upon to better understand the nature of the phenomenon, but small-scale laboratory experiments and reproducible observations are not really capable of touching the large-scale excesses that we are concerned with here. Historians could put more emphasis on regularities that can be observed in a comparative approach to the large-scale "experiments" that mankind is conducting, instead of just focusing on the peculiarities of the individual outbreak. Politicians could strive to put more energy and conviction behind international measures and organizations. It must not be forgotten, however, that politicians tend to be part of the problem rather than a solution, as the best way for a politician to build a constituency and power basis is to formulate some group interest and depict an enemy, that is, to engage the group instinct. A concrete consequence of this is that the UN is crippled in its peace-keeping task by the national interests of its member states. With our present level of understanding there is no simple marching plan to be enacted. We cannot turn pacifist and abolish the military, which will only invite aggressors to rise. We need to get a clear picture of how group violence arises to develop the equivalent of smoke detectors, and we need to fundamentally rethink international law to make it possible to intervene early enough. We need, above all, a clear conceptual framework to sort out the desirable from the ugly. Group strife may be necessary to keep mankind in shape and to make it progress, but then we have to define acceptable constraints for it. What the situation therefore most urgently cries out for is that human society takes note of the importance of the issue, and that an ongoing discussion process brings more light to it.

The Role of Cultural Constructs

Are we subject to a deep dilemma, having to suffer group violence as necessary evil, as the flip-side of culture, civilization and progress? No, we do have a choice! Our behavior is not shaped directly by our genetically controlled set of instincts. For the sake of simplicity and applicability to varying environments instincts have been formulated by evolution in abstract, schematic form. It needs a process of maturation and education to fill instincts with concrete structure relating to our actual environment and social situation. Thus, the archetypal formulations of who is part of our group and how to define strangers are filled with concrete substance by a process of imprinting, education and learning. This process is deeply influenced by

cultural tradition. Importantly here, the definition of what constitutes a valid group
—tribe, language community, religion, sect, nation etc.—is an issue of concepts
that are subject to historical change. Moreover, it is an issue not of genes but of
societal or historical events into which group activity we are drawn. Under the right
combination of factors, maybe under the influence of some social crisis and of a
gifted demagogue, we are all, with high likelihood, prone to be drawn into violent
group activity. We are not born monsters, we are made into monsters by social
mechanisms that are subject to cultural change.

If mankind wants to avoid these conflagrations, it has to work on understanding
the relevant mechanisms and shape the relevant cultural traditions. The structure of
these cultural traditions is not easily understood or influenced, but changing them is
our only chance at avoiding the bad instabilities of group violence before they arise.
Our genes have unavoidably endowed us with the group instinct in the form of an
abstract schema, but human society should be able to sculpt the way this instinct
unfolds in the educational process. The instinctual definition of who belongs to my
group and who doesn't is evidently based on characteristics perceived directly
during personal encounters (Kinzler et al. 2007) such as familiarity in aspect and
expression and the ability to communicate and to emotionally relate. These emotional mechanisms are routinely exploited to extend the boundaries of groups
beyond personal experience, with the help of common language, ritual and cultural
symbolism. This extension of the definition of groups beyond immediate personal
grasp creates larger units of coherent collaboration, but it brings its own grave
dangers. Building a nation avoids group violence between clans or provinces, but it
also creates a more potent group with more power to do violence, as history
teaches. The tool of demagogues to build their power base is to define and create
large groups, on such common traits as need or class or language, and by putting up
other groups as common enemies, and it is deplorable that mankind continues to
callowly fall for these attempts again and again. It is perhaps time to start building
cultural constructs, a world view—a picture of human existence, a Weltanschauung
—that takes into account the biology of human nature and social and cultural
mechanisms such as to avoid grand-scale violent conflagrations.

The Protagorean World View as Problem

Anything that goes beyond my personal social experience is based on mental
constructs. The presently dominant world view is problematic in not knowing its
own limits. Implicit in it is a set of four tenets that are widely held, though mostly
on a subconscious level, which makes them all the more powerful. I will call it here
the Protagorean world view and claim that it can, on final account, be blamed for
many of the disasters of human history. All four tenets are due to the tendency of
thinkers to take their own individual makeup as model of the world, "man as the
measure of all things," as Protagoras said.

Truth: There is no truth if it cannot be shown to me as individual. This tenet is central to the rational world view. Mathematical or experimental proof and visual evidence, are taken as the only means of establishing a fact as real. This tenet denies the existence of structures and patterns beyond the horizon of the individual.

Picture of Man: Mankind—human society—is pictured as statistical ensemble of individuals. This amounts to seeing ourselves as part of a structure-less mass. It is as if we wanted to conceive of a piece of music in terms of the statistics of its notes. This view denies the existence of overarching patterns and structures spanning many individuals. This tenet is a direct corollary of the previous one, in being the Procrustean attempt to cut off from man whatever goes beyond the individual perspective.

Values: Human values are individual values. The American Constitution puts up as national goal the pursuit of happiness, an individual value. This tenet declares the individual—essentially anybody's Ego—as the purpose of it all. From this perspective, all social structure can only be the result of the compromise of curtailing my own interests to give room to others.

Organization: The only organizing agent in this world is the individual human mind.

Pronounced like this, these four tenets will generally be rejected, but they nevertheless are, to a very large extent, the subconscious basis of our way to organize human affairs. The most fundamental of them, and the one held with greatest conviction, is the idea, which has gained the upper hand in the Age of Enlightenment, that there can be no truth—no reality—beyond the radius of individual insight. Our modern society is an extremely complex constellation of millions of individuals, and there are patterns that cannot be formulated in terms of concepts accessible to the individual. Each individual has detailed insight into a social context that comprises at best a few hundred individuals. These spheres of understanding and influence all overlap to form the coherent whole of society at large, but the global whole of this reality is beyond the individual's mental horizon. By analogy, thoughts in our brain are constituted as grand patterns of firing nerve cells, way beyond anything that can be formulated in terms of the sphere of influence of the individual neuron. Denying or ignoring the existence of the reality of the overarching structures constituted by the inter-meshing of the individual spheres of influence, by insisting that the individual can comprehend the whole of society, forces us to come up with grossly simplified primitive pictures of man. Further, if the individual can comprehend the whole of society, the fallacy continues, and if the individual mind is the only motive force in this world, then it is conceivable, or may even seem unavoidable, to give power to a single individual to run the whole and to define the structure and purpose of society.

It is true that in our society another world view is still alive, although rarely given clear expression. According to it, I as an individual see myself as part of an organism, which has its own higher goals and reasons, which is high above me and which I cannot hope to understand. My role in life is to struggle and find my place in this world, to conform to what is asked of me, to do my duty, to find fulfillment in being part of that higher organism. We hear such words spoken in church, but it is sadly true that religion itself to a very large extent is maimed by the Protagorean world view, formulating apodictic truths to be forced on others (Gray 2007), picturing a personal god in the image of the individual, a god who is responsible for creation, who is to be argued with in terms of utilitarian values, and who is misused to define groups and incite violence against non-believers.

By insisting that we understand the whole of society and that society has to be organized in its entirety by the individual mind—be it that of dictator, legislator or voter—we are bringing the cultural organism of society down to what we understand. The conflict between two groups of millions of people is a construction, a legend, made up by individual minds for individual minds. Wars and group violence are based on social constructs that claim to be the reality but which have very little to do with reality.

The situation is more complicated than painted so far, as there are global realities that *are* accessible to the individual, and which call for mechanisms involving individuals in power positions. Natural disasters or military aggression brought over your group are such realities, and they call for measures accessible to the individual mind. When, however, Hitler divided the world into superior and inferior races, and when Stalin divided society into historically winning and losing classes, and both of them made concrete predictions as to the course of future history, it was all mere confabulation: only the Protagorean world view gave them the power to turn their predictions into frightful reality for a period of terror. Without the pervasive belief that theories like that of Marx or Darwin could tell mankind's future or serve as action plan for breeding Ubemenschen these systems could not have gained a foothold in the minds of the masses.

What can be done? The Protagorean world view took centuries to dominate the West, and cannot be abolished in a stroke. Also, many truths in and about this world are indeed accessible to the individual, and many things can be organized by the individual, individual values have to play a role in our life, although not the prime role, and even society on its global level can be understood in some of its aspects by the individual. It will need a protracted process of sorting these things out and evolve a more realistic world view, essentially revoking the four tenets of the Protagorean world view. A first and very important step is the realization that the reality of man cannot be grasped by the individual. Tightly connected to that is the necessity to realize that for man to be organized there is no need for the individual to understand it all. Human society can shape itself with its own organismic mechanisms, of which we individuals are part and not master. The true historical forces that create and stabilize civilizations and political orders are not the play material of individual thinkers such as Marx or Hitler, but are processes that let lifestyles and world views grow over generations through organic mechanisms spanning whole societies. Individual thinkers and societal leaders are not unimportant in this process, but if they are guided by the megalomania of the Protagorean world view the result of their efforts can only by short-lived fashions or social disasters. We must free the process that shape our lifestyle and traditions from the disease of the individual mind taking itself too seriously so they can develop naturally. If we take to heart our own individual inability to understand and guide society globally and in detail, we will discover, or rather rediscover, the means and mechanisms with which we can turn ourselves into creative instruments in the hands of the super-organism of which we are part. We have, over the centuries, created powerful instruments of amplification of individual power, vast companies subject to quick decisions by a handful of managers, military hierarchies and legal and executive government structures all able to reach deep into the lives of every citizen. It is these power amplifiers that project our group instinct, our urge to do what our society expects from us and to follow a leader, onto a level and scale that is no longer human. Only if mankind realizes the truth that society as a whole cannot be understood in detail by the individual and that society is well capable of organizing itself by collective mechanisms, will it be ready to dismantle those power amplifiers. If a changed world view were to direct us to content ourselves with our direct social sphere of influence and let the rest to be taken care of by super-personal social forces, the edges would be taken out of the threat of group violence.

Fighting the phenomenon of large-scale group violence is not possible with concepts shaped around the individual and its actions. Threatening the instigators of group violence with personal punishment is a toothless measure. In the grips of mass movements, individuals don't feel the threat of punishment as substantial, and besides they are willing to take the risk. During the Nuremberg and Eichmann trials, it became evident to observers that without their power amplifiers those men were no longer the forces of evil that were to be hit. With the dismantling of Nazi Germany those forces had disappeared. It is rather ineffectual to educate the individual about atrocities and how to avoid them, history shows that group forces are stronger than individual convictions and that our group instinct is able to induce us, at least the majority of us, to do whatever our group orders us to do. We will have to enlist collective forces, forces that are high above the individual, to change the course of history for the better. Germany still feels bowed down by the conscience of the collective crime it committed and may be immune against repetition as long as this conscience is alive. But mankind as a whole is very far from the realization that our common biological nature, our group instinct, is ever-present fuel for dreadful conflagrations.

Summary

Group violence is an instability of human society that threatens to strike just anywhere. The neurobiological basis for it is what I call here our group instinct. This instinct is essential as social glue to let us collaborate harmoniously inside our group. However, it does not extend to members of other groups, against whom we can be incited to commit arbitrary acts of violence. It is important to raise worldwide consciousness that this instinct is in all of us and that in this sense we are all time bombs.

The original structure of the instinct is directed at the social environment we experience directly, but is lifted to large scale by social constructs such as nation, race, class or religion. These social constructs rest on the misconception that the individual mind is capable of detailed understanding of society as a whole. Overcoming this misconception will rob large-scale group violence of its breeding ground.

Policy Recommendations for Future Action:

- Establish a forum to focus worldwide attention on the group violence phenomenon.
- Organize a series of conferences bringing together scientists, thinkers and society leaders to discuss and study the issue of group violence and prepare concrete action.

References

Arendt, H. (1968). *The origins of Totalitarianism.*New York: Harvest, Harcourt, Brace and Jovanovich.

Bloom, H. (1977). *The Lucifer principle: A scientific expedition into the forces of history.* Atlantic Monthly Press.

Canetti, E. (1973). *Crowds and power.* Penguin Books.

Freud, S. (1975). *Group psychology and the analysis of the ego.* New York: Norton.

Gasset, J. O. Y. (1932). *The revolt of the masses.* New York: Norton.

Gray, J. (2007). *Black mass: Apocalyptic religion and the death of utopia.* Publisher: Allen Lane.

Haney, C., Banks, C., & Zimbardo, P. (1973). Interpersonal dynamics in a simulated prison. *International Journal of Criminology and Penology, 1,* 69 –97.

Kinzler, K. D., Dupoux, E., & Spelke, E. S. (2007). The native language of social cognition. *PNAS, 104,* 12577–12580.

Le Bon, G. (1952). *The Crowd.* London: Ernest Benn.

Milgram, S. (1963). Behavioral study of obedience. *Journal of Abnormal and Social Psychology, 67,* 371–378. (Lancaster, Pa).

To Be Different: How Medicine Contributes to Social Integration of the Disabled in the Era of Globalization

H. Binder

Studying epidemiological data and projections for the coming decades, one must assume a numerical increase in disabled persons suffering from diseases of the brain will occur. The reasons are manifold but the main responsibility must be borne by dramatically changing environmental and social living conditions and increasing life expectancy particularly in the developed countries, where most of the so-called age-dependent diseases arc to be found. Medicine, not least, is responsible for this situation because improvements in treatment for formerly fatal diseases now lead to the treatment of recently evolved chronic diseases. Looking through WHO publications, the proportion of the disability burden (YLD-Years Lost due to Disability) caused by diseases of the brain can be clearly detected on the basis of the YLD data for the year 2002. The number of YLD's attributable to neuropsychiatry—including cerebrovascular diseases—amounts to nearly one-third of the total YLD's. And this proportion will not improve; on the contrary there is much fear of it worsening. The reason why brain diseases produce disability is due to the fact that the brain is the highest morphological substrate for movement, cognition and language; and one must not forget, of course, feelings and lastly, human behavior. Discussing disability necessitates two opposing points of view: a self-description by the disabled person and another person's particular social point of view. Both depend on capability, verbal and soft skills and not least on the body as a whole.

These aforementioned categories pertain to all of the brain diseases. And apart from deformations caused by physical injury there are no other diseases apart from brain diseases which distinguish a person as "different". Basic neurosciences have made and will continue to make substantial progress not only as regards the pathophysiology of brain diseases but, also, how the brain works, which in turn imparts all that knowledge and those skills which are features of the individual.

H. Binder (✉)
Neurologic Center, SMZ Baumgartner Höhe Otto Wagner Spital,
Wien Baumgartner Hohe 1, 1145 Penzing, Austria
e-mail: heinrich.binder@wienkav.at

© Springer International Publishing AG, part of Springer Nature 2019
J. A. S. Kelso (ed.), *Learning To Live Together: Promoting Social Harmony*,
https://doi.org/10.1007/978-3-319-90659-1_6

Above all, the latter is essential in understanding how neurorehabilitation turns to good account in its association with disability. Alas, it gives the impression that medicine has backed away from the human being as a whole, to bits and pieces of cell—even perhaps, that of genome -and cedes responsibility to ever accumulating and expensive machinery/technology/equipment. As regards impairment, this might well be its future, in particular, the so-called context factors. But apart from who can or will afford it, this development is not the answer to the problems of activity and participation which matter to a person's self-portrait. In this regard, it is necessary for humanity's benefit, to change and shape society's attitude towards this problem. If it is true that a society will be judged by how it treats its most vulnerable members, the appropriate contact with the disabled; acceptance and care is badly needed. This is the main task society will be confronted with in the next decades.

What does it mean to be "disabled"? It means to have less freedom because of internal and/or external factors. Or, expressed in a different way, a person is regarded as disabled when he/she is insufficiently integrated within the multifactorial human-environmental system, because of failure or decrease in efficiency (Sanders 1997). We can therefore come to two basic starting points: a medical and a social one. The latter puts disability within a greater sphere and also contains the broader spectrum of the acknowledgement of the rights of persons who are disabled. Here, the definition of being disabled itself and what is regarded as being disabled, has its effect upon us all—how disabled persons are accepted by society in general and how they are seen and treated by administrations and other institutions (Definitionen des Begriffs "Behindcrung" in Europa 2002).

It is a great pity that the major part of literature on this topic deals only with bio-medical factors with little reference to its socio-cultural content (Groce and Zola 1993). Insofar as this is of great importance, it needs to be changed—as one can assume that about 80% of all disabled persons live in developing countries and of these, 60–70% of the disabled live in rural regions (Helander 1993).

Disability and Society

There are various kinds of disability which are also perceived in different ways from the people who are themselves disabled and on the other hand, by society itself. We find disablement not only because of physical failures, of behavior, of intellectual efficiency but also because of belonging to a certain social stratum, a religion or an ethnic group.

How an individual reacts towards disability depends on what the ideas and values of the society to which he belongs are—a negative attitude in childhood (parental prejudices for example) and then normal conflict between these negative attitudes and disapproval of showing openly these affective tendencies. Therefore ambivalent feelings generate pity, sympathy, impersonal help and finally, feigned acceptance. The approach towards disability is not the same in all cultures. It is therefore a great pity that comparative studies of the disabled in various cultures are less than one generation in existence and are almost completely confined to individual nations.

It appears that nearly all human societies categorize as Disabled, the existence of physical, psychical or also the sensory characteristic qualities which distinguish disabled persons from the non-disabled members of society. The cultural interpretation of this "being different" varies significantly from society to society and above all, societies differ in their attitude towards and evaluation of the various kinds of disability. This occurs principally according to three criteria:

(1) Causality: When the question as to the reason for any disability arises either from the disabled person himself or from the society to which he belongs, there are obviously two possible explanations. There is one explicit, logical and scientific answer but quite often also an implicit culturally-specific interpretation. The latter extends from "Act of God" -through a non-specific "higher power"—until it reaches "through one's own fault". The question as to cause was and is nearly always of importance for the competence of a given social institution.

(2) Kind of Disability: Various groups in a society have quite different either positive or regrettably, more often than not, rather negative attitudes towards physical and intellectual ability or the deficit of same. Therefore the integration of a disabled person depends largely on the standards and values of any society within a given culture.

(3) The role of the disabled person in society: This seems to be a question of the ethics of obtaining in any given society. It is mainly a question of the value a disabled person is to society. It largely depends on the willingness of society to disburse from its resources and how the role of the disabled person within that society and his future contribution to it is estimated. Regrettably, here also gender seems to play an important role.

The continuous development of modern civilization has not only changed the role of the individual himself, but also the role of the disabled person. In societies where most of its members are working manually, where the center of work lies in agriculture or a similar activity, it is probably not so difficult to find a way to engage people who are suffering from a light or medium degree of disability. On the contrary, countries with highly developed industries and technologies seem to prefer to keep disabled persons within a kind of pension system rather than integrate them into working processes. And this although also no less disabled, could care quite productively for their own livelihood as well as for others. Even when some persons might not be able to work outside the confines of their homes, they would still be able to contribute by their work, for their own and their family's upkeep. But even they are exposed to a certain kind of discrimination and they are only given jobs that others are not willing to perform. Finally, this leads to a state where, in certain civilizations, they have only the role of beggars in front of churches, temples, mosques, on car parks and railway stations, in order to scrape a living for themselves and their families. The more striking the disability the greater the chance of receiving alms.

But also in their own families the disabled are very often outcasts, find no support, and are regarded as a burden and are banned from public gaze and locked up in some back room and banished to the last corner of the flat or house. The conscience of relatives of disabled persons and of society in general is complicated by the fact that although basic medical treatment, also the need of food, clothing and lodging are met, in actual fact, the disabled person is excluded from participation in family or social activities.

The attitude towards disabled persons has in these last years, at least in the industrialized countries, changed somewhat. At least, selectively, they are no longer hidden away. For this, the media and sport have contributed much. Although the public still restrained, the openly admitted Alzheimer's disease of the late President Regan and movies like "Rain Man" have aroused attention to some forms of intellectual disability, "Rain Man" is especially a good example of how one can deal with the problem of wider acceptancy and not only show the darker side of disability. Regrettably, one can also see perversions, as in the movie "Planet Terror" where a submachine gun replaces a leg prosthesis!

The sports public attention towards the disabled is headed by such events as Paraolympics and one can admire their abilities via various media. It is really surprising that disability can be turned to advantage because of the developments in modern techniques of ortheses, or, can at least, be regarded as such. Remarkable in every way is the case of the South African Pistorius, who had both lower legs amputated. He participated as a "disabled person" at a normal light-athletic meeting and came second in a 400 m event in his carbon prosthesis. He then applied for participation in the Olympics in Rome. His application was rejected by the IAAF which reasoned that he would have an advantage over a non-disabled person by virtue of his prosthesis!

The question arises, whether the normal God-given human body will soon no longer be good enough. How long will it eventually be that no longer will aesthetic improvements be carried out on the human body, but we will also try to improve efficiency by replacing parts of the bodies of healthy persons? First attempts can already be registered in the form or the so-called neuroenhancement, which tries to increase, using drugs, the ability of students to both learn and fit them better to the demands of their studies. Representatives of enhancement technologies are already arguing that through progress in neurotechnology, one can also increasingly improve the efficiency of normal healthy persons. Why should one not give a person the chance to achieve a higher grade of concentration, when such neurotechnological operations (seemingly?) have no side effects?

The Identity of the Disabled

Each individual perceives himself as a member of the various groups to which he belongs (Sen 2006). Each of these groups gives him a certain identity. As regards this unavoidable plural identity, one has to decide, when a certain context is given, what kind of relevance has to be attributed to the various bonds and affiliation. The problem will arise, if one discovers it, that there is unavoidably one and only one identity, which apparently demands very much from the single individual as well as from the accompanying society.

Regrettably, the choice of identity is only possible within the limits that we find achievable. The practicability, as far as identity is concerned, depends on individual characteristics and circumstances, which define the possibilities that are open to us. The problem of disability now is, that this identity is not freely chosen, but imposed and that there is no alternative existing. This goes together with rather strong and exclusive feelings of belonging to a certain group—in this case, the one of the disabled—linked with the feeling or acknowledgement of distance from or divergence from other groups, especially from the group of the healthy.

Disability and the Burden of Disease

According to an estimation of WHO, about a billion people are suffering from a neurological disease. Beginning with dementia, through diseases of the brain vessels and epilepsy until headaches. Some of these diseases or of the ailments which result from them, are at the present time, impossible or only very difficult to treat. That is the case of Alzheimer's disease, the number of sufferers from which will be a fast growing one in all industrialized nations (Newton 2007). In developing countries on the other hand, many neurological diseases are highly stigmatized. Besides this, one must not forget prevention as concerns risk factors of stroke, injuries to the brain because of traffic accident or poliomyelitis (Editoral 2007a).

During these last years, one tried very hard to find internationally key–numbers for the state of health for certain defined groups of people. If one reduces the findings only to the numbers for mortality, one sees that the significance of neurological diseases is certainly underestimated. A glance at mortality rates from dementia makes this abundantly clear—the number lies, world-wide—under 1%. We can illustrate this even more starkly by referring to the DISABILITY ADJUSTED LIFE YEAR or DALY (Field and Gold 1998; Murray et al. 2002) which is a health gap measure that extends the concept of potential years of life lost due to premature death (PYLL) to' include equivalent years of "healthy" life lost by virtue of being in states of poor health or disability. The DALY combines in one measure the time lived with disability and the time lost due to premature mortality. One DALY can be thought of as one lost year of "healthy" life and the burden of disease as a measurement of the gap between current health status and an ideal situation where everyone lives into old age free of disease and disability.

According to DALY encumbrances through neurological diseases by far exceed most of all other diseases, with the exception of terminal cancer and injuries of the spinal cord. In 2005, more than 92 million healthy life years were lost through neurological diseases, more than half of these through diseases of the brain vessels. In the industrialized countries and in the developing countries, we find stroke as number 3 in ranking of all the important burdensome illnesses. But in fact, we have 6.4% stroke in industrialized countries and only 5% in the developing countries, according to DALY's. Only in developing countries with a high mortality rate, stroke ranks as number 10. But because of very good acute care and rehabilitation for stroke patients in industrialized countries, only one-fifth suffers later on from a more serious kind of disability, whilst in the developing countries at least *2/3* of the patients need support in order to come up to their daily basic needs, a support which they regrettably never get (Southern African Stroke Prevention Initiative 2004; Hayward 2004). Over 11 million DALY's fall victim to Alzheimer's disease as well as other kinds of dementia. And Epilepsy and Migraine were responsible for another 15 million DALY's. The basic data from these findings come nearly all from industrialized countries. The fact that the burden of neurological diseases is so comparatively low in developing countries compared with high-income countries, lies in the non-validation of data and in lower life expectancy. But we certainly will have to expect an increase (Editoral 2007b).

Violent crime, including suicide attempts, war, etc. are increasingly responsible for the burden of illness among young, economically productive grown-ups. Whilst in the industrialized countries accidents, attempts at suicide and suicide itself, are the major problem, in the developing countries violence and war preponderate. The latter is also valid for the former Soviet Union and other countries having a high mortality rate, where the frequency of death due to accidents and disability is similar to the number in the Subsaharan region. Regrettably, those hit by the most severe and mutilating injuries are, in first line, young grown-ups. Worldwide, injuries were responsible for 16% of the burden of illnesses in the year 2002. In parts of America, Eastern Europe and the eastern Mediterranean, more than 30% among the male grown-ups between 15 and 44 years, fell into this category. In the age group from 15 until 44 of the male grown-ups, accidents, violence and suicide are among the 10 most frequent reasons for the burden of illness. Globally viewed, in 2002, the third most frequent reason for the burden of illness was topped only by HIV and depression. The rate of traffic accidents is increasing significantly, especially in the developing countries, the Subsaharan region, South Asia, and Southeast Asia, and concerns men in particular.

Future Development—Population

There is no doubt that the population of the world will increase dramatically in the next decades, in the course of which all scenarios predict about 8 billion people by the year 2030. Though expansion will be different in the various parts of the globe, one can assume that the population of Europe will slightly decrease and the number of its people will, by 2030, lie between 820–830 million, whilst it will increase in Southeast Asia to more than 2 billion and in Africa it will increase to about 1.1–1.2 billion.

Increase in population in the cities also means increasing problems. The number of city-dwellers in the developing countries will increase by more than 3% per year; that is, triple the increase in numbers of the rural population. That means that in the developing countries, the number of people will increase by 100 million persons per year, because of migration to the cities and increasing birth rate. Hence, the number of city dwellers will exceed the number of people in rural regions. Already, one out of three city residents, all told, about one billion, lives in slums under circumstances which are determined by bad air, little care, lack of security, lack of drinking water, bad sanitation, poor health care and violence, especially against women. It is therefore not surprising that illness, high mortality and unemployment make their appearance in the slums comparatively much more frequently than in other urban areas. Bad housing conditions, lack of sanitation, air pollution, noise disturbance… the list of reasons responsible for people becoming ill is a long one.

Rudolf Virchow already in 1848 realized the correlation between health and people's sociopolitical environment: "medicine is a social science and politics is medicine in a broader spectrum." In this instance, one finds initial attempts in city planning which included public health aspects. Already half a century later, in 1875, Sir Benjamin Ward-Richardson described utopian towns with clean air, good public transport, small hospital units in the city quarters homes for the elderly and disabled, cities without tobacco and alcohol, which would be safe and healthy for everyone.

Future Development—Disability

The disease burden of the so-called non-communicable diseases increases with the years and is already now responsible for more than half of the worldwide burden of diseases at all stages of age. One can assume that there has been an increase of about 10% since 1990. Whilst the share of the non-communicable diseases among grown-ups in industrialized countries lies relatively stable at 85%, the rate in countries with a medium income has increased by about 70%.

When one compares the year 1990 with the prognosis for 2020, cerebrovascular diseases—and therefore stroke and its consequences -will have progressed from 6 to 4th place in ranking. One also assumes that traffic accidents will change from 9th to 3rd place! One considers that the percentage of the burden caused by non-communicable diseases in present developing countries will drastically increase through longer life expectancy and a different distribution of risk factors.

Cardio-vascular diseases arc responsible for 13% of the burden of diseases among grown-ups and infarct of the heart and stroke are the two leading reasons for mortality and burden if disease in the group of working people above the age of 60. They are also among the first 10 reasons in the age–group between 15 and 50 years. In the industrialized countries diseases of the heart and cerebro-vascular disease are responsible, commonly, for 36% of the mortality whereby the death rate is higher for men than for women. The increase of cardio-vascular mortality in Eastern Europe runs in contrast to the continuing decrease in other industrialized countries. Mortality and cardio-vascular conditioned burden of disease is massively increasing in the developing countries.

What Can Medicine Do for Disabled People?

Regrettably, about the turn of the century (19–20th centuries), a new kind of medicine appeared. For nearly one century, natural-scientific medicine was believed to be a possible victor over all diseases. This victory was not possible as medicine understood itself in a nearly only curative way, as a weapon of war against ever newly out breaking diseases. In the meantime all of us should be aware that medicine can neither guarantee a stable or a return to health, nor can it keep ageing and death from us (Illich 1975). There is no such thing as a clinical, social and cultural Iatrogenesis. But medicine can help in the prevention of disease, it can cure many but certainly not all diseases, it can help to restore functions of the body, of organs, and by doing so support the patient's social life; the latter would already have a sociopolitical aspect, such as Virchow has already discovered in the 19th century with his hint of a healthy environment.

Medicine should never lose its holistic view, as Vichow has seen it, neither in research or in intercourse with patients. One should never forget what people really want and need. Medicine has to act in accordance with ethical principles when applying its curative and rehabilitation methods and not exclude anyone from them. Therefore it is absolutely necessary to change medicine's outlook according to the Copernican concept. Medicine should not orientate itself from deficits in efficiency but concentrate on what there still exists and preserve and support it. This is the first and central part of rehabilitation, apart from prevention and curative medicine. The sphere of neurohabilitation becomes more and more important especially when seen in relationship to the consequences of neurological diseases.

There is no other disease -if one omits rather mutilating injuries -which makes a person different from others, than the stigmatizing diseases of the central nervous system, inclusive of diseases of the psyche. In the last decades the results of neuroscience were mostly ignored by the broad public, but since the eighties of the last century attention has increased. Neuroscience has greatly helped in the understanding of the functions of the nervous system. It has developed therapies which help to ameliorate the consequences of diseases of somatic ·and intellectual disabilities. As far as drugs are concerned, many and more powerful drugs against pain, epilepsy, spasms and many others, belong to this group. Great progress has occurred in the last ten years in the field of medical technique. Today, one can influence in a positive way such elements as pain, movement disorders, epileptic attacks and even affective disorders like depression, by using implants and electric impulses in the peripheral nerves, in the spinal cord and even in the brain. There even exists so-called "exoskeletons" which, as in science fiction movies, support the motoric system. Research in stem cells should also be mentioned, but it is still far from being applied generally.

But all these findings and inventions cannot replace a human being who, in an understanding and delicate way, takes the hand of the disabled and guides him out of his isolation and leads him from his categorized little box, labeled "disabled", back to society and its social context. This is the central task of rehabilitation; and because this is so, it needs sufficient and specially-trained and competent personnel resources.

Globalization and Its Effects

Globalization and increasing solidarity between countries, the opening of borders to ideas, persons, economies and financial capital has positive but also stressing effects on people's health.

One of the negative effects existing for some years is the migration of medical personnel. In the past, the migration of well-trained medical personnel from poorer to richer countries was a passive process, mainly propelled by the political, economical, social and professional circumstances of each single migrant. But in the last years the demand for medical personnel has increased in many of the western industrialized countries, above all because of the dynamics in the changes of population. In many of these western countries one is meanwhile dependent on the immigration of medical personnel, while on the other hand, it has catastrophic effects on the health system in developing countries, especially Africa.

If one analyses the reasons why experienced health personnel emigrates, one can find two major categories: the push and on the other hand, the pull factors. The push factors are: low income, poor motivation, lack of supervision, limited chance of promotion, obsolete equipment, lack of basic medical supplies, dangerous conditions of work and last but not least transgressions of human rights, ethnic, religious and political persecution and war. The pull factors also include economical reasons, but also the chance of professional development and safety in the work place.

What has gone wrong? The changes in the population dynamic in the industrialized countries have produced an ageing population with an increasing demand for health services (Organization for Economic Co-operation and development 2002). The reasons for this are a low fertility rate, longer life-expectancy, early retirement, overageing of the working population. All this is connected with constantly decreasing numbers of health personnel especially in the nursing domain. America, for instances, expects a deficit in the numbers of nurses of about 500,000 by the year 2015. The goal should be to cover the legitimate needs of the industrialized countries without neglecting the health system requirements of the developing countries.

Another burdening effect is the rapid progress in possible medical treatment which costs more and more money but at the same time is far from basic needs and only an increasingly shrinking social strata can afford it. Undeniable success in medicine is also the success of medical industries. Medicine needs money from industry and by the same token, industry is served by medicine. The development of medicine and drugs is rather expensive. The industrialized countries have to invest in increasingly more and more money for outlays for safety, court costs, etc. Companies have to expand and amalgamate with transnational units. The shareholders demand more dividends. Consequently, many work places are outsourced to countries with cheap available labor or are reduced in number and trials are transferred to developing countries. This increases participant risk but brings more profit at least for a certain time. Finally, the successful drug will be sold in the industrialized country. The gap in the already existing 2–class medicine broadens (Berlinguer 2004).

Although everybody agrees that health belongs to basic human rights and all human beings should have the same and free access to a basic provision of health services, for the poor it becomes more and more difficult to avail themselves of such service. The reasons are mostly financial in nature with governments trying to push costs from the state to the private sector. Therefore, expensive drugs, expensive hospital stays, long-time therapy, the latter including rehabilitation, are no longer available in the poorer countries and have also to be rationalized in the so-called rich countries. Consequences are sickness and disability.

In the past attention was given, in first line, to the control of infectious diseases, national security deficiencies, the supply of affordable drugs and changes to international goods traffic and diverse financial deals through which a better access to medical treatment was made possible. Regrettably, one has overlooked that the landscape of diseases has changed, that heart illnesses, stroke, diabetes, overweight and injuries in general are increasing. In the meantime all these illnesses become more and more relevant in the global pattern as well in mortality as in disability.

Also in the meantime, health -and above all, social policies -acting as a compensatory limit to capital and industry in conditions of border-free finance -as well as workers' mobility -become more difficult. Among other things, this is shown to be so in the succession of developments in a global market for socio-political moves involving clinic managers, insurance companies, etc. also competition between welfare states and the appearance of new actors like "Attack" or the World Social Forum in Porto Allegre (Deacon 2003).

The WHO realizes the importance of and the increasing number of disabled people and also the necessity for better provisions and rehabilitation for them. Not only because of chronic diseases but also of population growth, the medical progress, which preserves and prolongs life. They refer to the Declaration of Alma Ata where rehabilitation is defined as one of the basic provisions. In this Declaration there is also stated that personnel in the medical field need training in the integration of rehabilitation into their activities, that disabled persons have to be integrated into the Training Staff and that there should exist constant communication between disabled, health politicians and society. In the meantime, also the big international organizations like World Bank, WHO and UNESCO now realize the far-reaching consequences that economic change has on health (WHO 2002; World Health Organization 2005). Regrettably, they can only suggest solutions. For the execution and supervision of these solutions, they lack the executive power and the power to carry it through against political and/or economic interests.

Summary

Medicine can contribute to social integration when by means of rehabilitation the disabled are helped to attain the necessary physical and intellectual requirements for such integration. Medicine has to accompany the disabled in the process of integration and help them avoid excluding themselves. This is possible only when political and social frame conditions do not hinder but support integration. Global discussion—in the meantime this has begun—on the very best social policies to be adopted has to be processed. One has to strive for a globalization with a human face (Townsend and Gordon 2002).

References

Berlinguer, G. (2004). Bioethics, health, and inequalities. *Lancet, 364,* 1086–1091.
Deacon, B. (2003). Global Governance reform: From Institutions and Policies to Network Projects and Partnerships. In: B. Deacon, E. Ollila, M. Koivusalo, P. Stubs (Eds.), *Global social governance. Themes and prospects.* Helsinki: Ministry of Foreign Affairs of Finland.
Definitionen des Begriffs "Behindcrung" in Europa: Eine vergleichendo: Analyse. Europaische Kommission. Genera!direktion Beschl1ftigung und Soziales, Referat *E/4,* 2002.
Disability, including prevention, management and rehabilitation. Report by the secretarial. World Health Organization, fifty-eight world health assembly, Provisional agenda item 13.14. April 2005.
Editoral (2007a) Neurological disease: time to reassess (Vol. 369). http://wwwthelancet.com. March Hl.
Editoral (2007b) Neurology on the global health agenda (Vol. 6). http://neurology.thelancet.com. (April 2007).
Field, M. J., Gold, G. M. (1998). *Summarizing population health: Directions for the development and application of population metrics.* Institute of Medicine, Washington, D.C: National Academy Press.
Groce, N., & Zola, I. K. (1993). Multiculturalism, chronic illness and disability. *Pediatrics, 91,* 1048–1055.

Hayward, P. (2004). Stroke disability in South Africa matches more affluent nations. *The Lancet Neurology, 3,* 261. (May 2004).

Helander, E. (1993). *Prejudice and dignity: An introduction to community-based rehabilitation.* NY: United Nations Development Program.

Illich, I. (1975). Limits to medicine: medical nemesis.

Jr, Newton. (2007). The oldest and the disabled-the care priorities. *Gerodontology, 24*(1), 2.

Murray, C. J. L., Salomon, J. A., Mathers, C. D., & Lopez, A. D. (2002). *Summary measures of population health: concepts, ethics, measurement and applications.* Geneva: WHO.

Organization for Economic Co-operation and development. (2002). *International Mobility of the high skilled.* Paris: OECD.

Sanders, A. (1997). Behinderungsbegriffe und ihre Konsequenzen ruT die Integration. In: Eberwein, Hans (Hg.): Handh\lch IntegrationSJll1dagogik. Kinder mit und ohne Behinderung lemen gemeinsam. Weinheim und Basel 1997, 99–107.

Sen, Amaryta. (2006). *Identity and violence: The illusion of destined.* New York, London: W.W. Nonon & Company.

Southern African Stroke Prevention Initiative. (2004). SASPI. *Stroke, 35,* 627–632.

Townsend, P. & Gordon, D. (Eds.) (2002). *World poverty. New politics to defeat an old enemy.* Bristol: Policy Press.

WHO. The world report 2002. Geneva: World Health Organization, 2002.

Systems of the Brain Responsible for the Herd Mentality and the Acceptability of Diversity May Determine If We Can All Live Together

H. Weinberg

The developing technology for an understanding of distributed, dynamic and interacting systems of the brain will be able identify the unique characteristic of each individual. This has significant implications for the enhancement of plasticity of the brain and for reduction of the herd mentality, and for the future of the human species in respect to the acceptance of diversity, and may be the beginning of the way in which we can All Live Together.

Herd Mentality

One of systems of the brain which is clearly characteristic of human behaviour, and presumably of all living systems that have a brain, is the Herd Mentality. And the other system, which initially appears to be contradictory to the herd mentality, is the Diversity of Brain Function, the reality that each brain is different, and that these differences are responsible for the individuality of each person. From an initial perspective there seems to be competing systems in the human brain systems: the herd mentality and the diversity of individuality. The question then is: How can diversity and the herd mentality function cooperatively—and—if there is an

H. Weinberg (✉)
Simon Fraser University, Burnaby, BC, Canada
e-mail: halweinberg.ca@gmail.com

© Springer International Publishing AG, part of Springer Nature 2019
J. A. S. Kelso (ed.), *Learning To Live Together: Promoting Social Harmony*,
https://doi.org/10.1007/978-3-319-90659-1_7

evolving technology to change the brain, in order to modify these tendencies, should we proceed with that?

The Herd Mentality: Humans seem to mimic behaviours similar to a flock of herding animals, and may not realize that their decisions and actions are largely based upon the requirement to follow what their 'herd' is doing. The German philosopher Friedrich Nietzsche was, I think, the first to critique what he referred to as the "herd instinct", in any human society. Modern psychological and economic research has identified herd mentality in humans as an explanation of why large numbers of people may act in the same way at the same time. Each herd includes a leader and the herd follows that leader.

The modification of this instinct, i.e., modification of the brain systems responsible for herding, could be implemented in the future, and it might be possible for the herd mentality to be modified such that, what seems to be a characteristic of the herd mentality, the acceptability for killing members of a different herd.

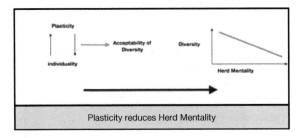

However, what would be the consequences if one were to eliminate the herd mentality in its entirety—the consequences of having no leaders, and no herds? Would humans survive and develop if there were no goals in which groups of humans participated, and if there were no leaders? How can individuality and herd mentality work together? One solution to control negative contributions of the herd mentality is to increase plasticity of the brain and consequently the acceptability of diversity,—a critical element in the determination of the individuality of the individual. Therefore, can the recognition, facilitation, understanding, and acceptability of the plasticity of brain systems as a method for establishing diversity, i.e., of continuously changing, distributed and interacting systems—of the diversity of brains, positively modify the herd mentality?

The history of brain physiology in the last 100 years is actually the history of a pendulum swinging, from the concept of localized function, to the concept of distributed function, through physically distributed, interacting and dynamic systems.

Until recently individual differences in brain function were described primarily in terms of deviations from the average, from a central tendency. However, recently it has been possible to begin an understanding of how systems in the brain change as the result of experience, how these systems are responsible for different people being different people. The identification and understanding of brain systems related to input, output, and neuroplasticity has expanded significantly as the result of developing technologies like Magnetoencephalography (MEG), Functional Magnetic Resonance Imaging (FMRI) and Positron Emission Tomography (PET) —and the use of non invasive brain stimulation (NBS).

Because of their time resolution, and their ability to directly measure the function of interacting neuronal systems in real time these technologies, and others, can be used to measure individual differences in the brain. These, and other methods, like non-invasive brain stimulation (NBS) using magnetic fields, and an increasingly advanced technology of brain imaging, will be able to recognize and document individual differences—and predict how those differences may impact the current and the future life of an individual, including the interaction of that individual with others, that is, the degree of their participation in the herd. An understanding of distributed, dynamic and interacting systems of the brain will be able identify the characteristic of individuals—and this has significant implications for the future of the human species in respect to the acceptance of diversity.

The history of attempts to understand the individuality of brain function dates back to 1700 B.C when Aristotle thought the heart, not the brain, was the location of intelligence for all individuals. Ancient Egyptians were probably the first to distribute written accounts of anatomy of the brain, and that anatomy was considered to be the same in all individuals, and the primary determinant of function in the brain. There was very little understanding of the neurophysiological basis of individual differences. As technology progressed, specific anatomical structures of the brain were identified as being responsible for different types of complex behaviour for all individuals.

The history of brain physiology in the last 100 years is actually the history of a movement from the concept of localized function—primarily but not exclusively for cortical function—to the concept of continually changing, distributed, interacting and dynamic systems. The early concepts of specific localized functions attributable to specific areas of the brain, has a long history that includes studies of brain stimulation and lesions of the brain including the use of frontal cortex lobotomies to treat behavioural disorders.

Frontal lobotomies, temporal lobectomies and other massive lesions were justified based on changes in behaviour that resulted from those lesions, without any clear understanding of how the brain functioned in respect to those behaviours.

In the late 50's I was doing "cutting edge research", i.e., the cutting of medial and lateral hypothalamus to study hunger, thirst and sex (it seems as if hunger, thirst and sex have always been related to each other, in one way or another). At that time, in the 50's, I wondered, to myself of course, whether a complex function like hunger, or sex, could be attributed to only one part of the brain, when the history of each individual clearly determined, to a large extent, the way in which those functions were individual to the individual, and until relatively recently, brain imaging of activity was primarily thought of as a method for measuring only localized function.

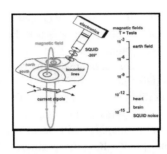

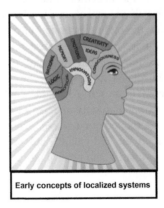

The first MEG to move around the head

An example was the first measurement of magnetic fields related to brain function, initiated by David Cohen in 1968. The recording was done using a copper induction coil as the detector. The idea was that electrical currents produce orthogonally oriented magnetic fields, and the net currents can be modelled as a current dipole, with a specific function, a location, orientation, and magnitude in the brain. This was, I think, the beginning of the idea to use "dipole sources" in the brain to explain different types of information processing. The SQUID was then introduced which used Josephson Junctions to detect very small magnetic fields, the Superconducting Quantum Interference Device.

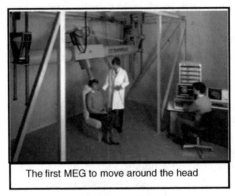

Early concepts of localized systems

MEG began with the assumption that it was a method for detecting dipoles. Dipoles were originally defined as an electrical source with a particular spatial orientation. The idea was that the dipole, located in a specific place, was responsible for complex information processing and response output.

Thus, the idea of a single dipole came to be the model of the sources of complex behavior. Of course an argument at that time was, and still is, whether the dipole is the location of "a source", or rather is an index of a complex distributed system. Later, it became increasing clear that the dipole was not a viable explanation of brain function responsible for complex behaviour.

The dipole was then considered as the 'centre of gravity' of a system, and was primarily considered an estimate, or average, of multiple sources, with respect to both direction and strength.

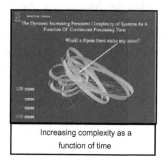

Increasing complexity as a
function of time

Brain imaging began to ask the question as to whether a dipole could be responsible for complex processing and an emphasis on understanding the distribution of function began to emerge. When I was working at the Burden Neurological Institute with Grey Walter, in Bristol England, during the late 60's, a period of social revolution within the Western world, we were doing multifocal stimulation of frontal white, with 64 gold electrodes that were implanted for as long as 6 months. This was to treat obsessive compulsive disorders. We worked from the midline outward to lateral structures, and observed behavioural changes specific to individual obsessions (Weinberg et al. 1969).

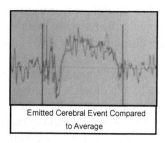

Emitted Cerebral Event Compared
to Average

We thought that we were reorganizing the interaction of subcortical and cortical systems, facilitating plasticity, but of course we really did not know what those systems were. We were doing what most scientists did at that time, and what is being done today to a large extent, i.e., trying out a treatment and determining its effectiveness, through an observation of its effect on behaviour. Although our knowledge of distributed function was limited we all began to realize that we did not have a real understanding of complex distributed brain function, and we hoped to get insight into the nature of those systems through studies of the distribution of evoked and spontaneous activity. In an attempt to study complex systems that were not the immediate result of stimulation we developed the concept of "emitted cerebral events"—events in the brain that reflected brain activity related to specific events when the events were not present—memories. The idea was to develop, for each individual, a template of the activity that resulted from the presentation of stimuli that required a response, and then to use that template for a match to what was happening in the brain when the patient was thinking about an expected but absent stimulus—it was called the Recognition Index. Of course, one thing we found is that the 'template' was different in different parts of the brain, and for different people—but there was a template that reflected what that patient was thinking about. Basically the endeavour was to use pattern recognition (Weinberg 1972) as an alternative to signal averaging. The idea we pursued in those early days was that an event in the brain was an electrophysiological change that could be described as a "template" for a "thought", a distributed system that could be discovered through the measurement of what was going on in a distribution of electrodes, over the whole head, when a particular type of information was being processed (Weinberg and Cooper 1972). The reason I mention this now is that it is an example of a methodology that attempted to describe complex distributed brain function, using spontaneous activity, related to information processing—activity that was different for each individual. We published some of this in the early 70's with Grey Walter, Ray Cooper, and with Rosa Gombi who was visiting from what was then the Soviet Union. However, the really important element of this research was that, for us, it began the attempt to identify patterns of brain activity related to information processing that was specific to individuals. The concept was that measurement of brain activity related to the processing of the same information may be different for different individuals.

I would like to expand a bit on this. I often think of Mozart. Can you imagine how a pianist can remember 10 different concerti—or more—and produce the frequently varied motor output related to those memories, i.e., to produce the same auditory concept? Clearly a distributed programme must exist that includes the use of sensory and motor systems, as well as the complex processing, and memory that occurs when each performance is almost, but not identical to, the last. And of course each musician could have a different pattern of brain activity that results in the same output, and that output is the characteristic of "that individual". Or think about something "a lot simpler". A person walking from point 'a' to point 'b'. Each person has their own gait, which results from their own input, processing and output systems, although they "perform the same action." As we all know you can identify someone by their gait: if one of those legs was amputated would you identify "walking" for that person as being located in that leg?

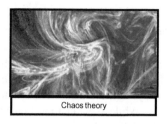

Chaos theory

The pendulum was clearly shifting to distributed systems in the late 90's. For example, Paul Nunez's book, Neocortical Dynamics (1995), and Gerald Edelman's book, The Remembered Present (1989), were important in the re-development of ideas of distributed function. Edelman described what he called re-entrant neural networks that were distributed systems that included the interaction of cortical, thalamic and brain stem activity. And then there was the application of Chaos Theory as another example (Skarda and Freeman 1990). Chaos theory applied to brain function considers the brain a complex, high dimensional dynamic system of billions of interacting sub-systems. The underlying idea of using chaos theory to study brain function is that complex function requires an interaction of widespread, spatially distributed functions. The assumption is that everything within the system is interacting. This is illustrated by the Butterfly Effect, whereby a single butterfly flapping its wings, akin to a change in a system of the brain as a result of experience, causes a tornado in the rest of the brain.

And then there was the application of nonlinear dynamical systems theory by Freeman and Skarda: "We have coined the term 'cooperative neural mass' to express this level of neuronal functioning. It is largely thanks to the analytical tools of nonlinear dynamics that we have been able to measure and interpret these spatially extended patterns of activity in the nervous system. Our approach has been to record and measure the neural activity patterns within the olfactory bulb before, and again after a subject had learned to discriminate two or more sensory stimuli, in order to identify precisely the differences in activity patterns that serve to distinguish and classify the neural events with respect to the discriminanda" (Skarda and Freeman 1990).

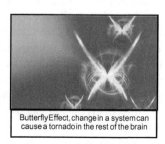

Butterfly Effect, change in a system can cause a tornado in the rest of the brain

In the early 90's emerged the NeuroChaos Solutions technology, itself based on Chaos Theory, a scientific principle describing the unpredictability of systems. Its premise is that systems sometimes reside in chaos, generating energy but without any predictability or direction. They are highly sensitive to initial conditions as illustrated by the Butterfly Effect mentioned above. Examples of these systems include the earth's weather system, the behavior of water boiling on a stove or the migratory patterns of birds. While their chaotic behavior may appear random at first, chaotic systems can be defined by a mathematical formula, and they are not without order or finite boundaries. Such systems consider the brain a complex, high dimensional biodynamic system of billions of interacting elements, capable of producing "Deterministic Chaos".

And then of course the concept of brain plasticity emerged as a focus—that there are changes in the brain's structure and function as a result of experience. The whole concept of plasticity includes the assumption that fixed and unchanging localization is not viable, and of course, the importance of "neuronal plasticity" is not only with regard to morphological changes in brain areas, but also for alterations in neuronal networks, including changes in neuronal connectivity and the generation of new neurons (neurogenesis), and also neurobiochemical changes.

And now there is emerging an emphasis on brain plasticity as a critical element in determination of the diversity of individuals. Brain plasticity is now clearly recognized as normal brain function related to the acquisition of behaviour and information processing—and the utilization of that information. Plasticity is now recognized as a fundamental property of the brain and has been implicated in various psychiatric and neurodegenerative disorders including obsession, depression, compulsion and psychosocial stress.

Plasticity is clearly a characterization of the individual as an individual—it may be that plasticity actually defines the character of the individual—and it is now clear that plasticity has become a real challenge to the concept of anatomically localized sources in the brain.

How can plasticity of the brain be facilitated? It is clear that the complexity of the brain's response to input, and to the organization for output begins at an early age, and is facilitated by variability of experience which has an influence on fluid intelligence, processing speeds, and the utilization of cognitive abilities (Kievita et al. 2016). Variability of experience in early ages and the practice and abilities for multi-tasking are of primary importance in development of plasticity. The ability to rapidly change attention and preparation for output, and to evaluate and deal with distractors varies between individuals and training. Experience with multitasking changes brain systems to improve performance of multitasking as a result of changes in the interaction of systems in the brain—the developing brain in a multitasking world (Rothbart and Posner 2015).

In the 70's, after coming back from the UK, where we were implanting electrodes in humans for long term stimulation to treat obsessive compulsive disorders, it was clear that the stimulation influenced some form of plasticity that resulted in the influence of memories on current systems of the brain but what was needed was a better method of recording the distributed function of dynamic systems.

Now back to the Brain Imaging, in the context of distributed dynamic systems that are unique to individuals—and the implications of this for the character of the human species. Magnetoencephalography (MEG) introduced of a new approach to the analysis of complex information processing because of its time resolution, and its ability to directly measure function in real time, without the use of any high frequency or chemical impositions on that function for the measurement of blood-oxygen-level dependent contrast imaging. The approach of MEG, and other imaging technologies was initially focused on finding source localizations that were part of the 'common brain'. When I returned from the Burden Neurological Institute, Max Burbank and his group were developing a single channel MEG system, and we began to collaborate in the development of studies of distributed systems—and at that time by multiple recordings using an MEG that mechanically moved around the head.

Of course at the time the idea was that there were fixed systems for processing input which could be identified by the configuration of dipoles that were computed, using different locations of the sensor. The CTF MEG technology began its development in 1970. In those early years the underlying SQUID technology was originally employed in geophysical exploration—at that time everyone was initially looking for dipoles in the brain. I remember asking if MEG could discriminate between excitatory and inhibitory systems. Inhibitory systems are of course critical in an understanding of distributed interacting systems and in the 70's I organized several symposia through the Canadian Psychological Association, to discuss the question of whether the contingent negative variation (CNV) was a unitary potential.

Current technologies are now focusing on the new look in brain imaging, i.e., the focus on Individual Differences in Brain Systems—and the control and modification of individual capabilities. For an understanding of individuals it is time to stop describing an individual with respect to the central tendencies, and standard deviations. Each brain is different and that is why we are different people. What will be the result of new technologies that will be able to control those individual properties of systems in the brain?

One of the new efforts to understand individual differences, and the use of those differences, is the consortium of Washington University, University of Minnesota, and Oxford University to begin comprehensively mapping human brain circuitry in a target number of 1200 healthy adults using cutting-edge methods of noninvasive neuroimaging—to understand brain connectivity, its relationship to individual differences in brain circuitry and to behavior—the Connectome Project (https://humanconnectome.org). From current studies the apparent primary goal of the Human Connectome Project appears to be an understanding of individual differences—differences that actually define the individual—the personal characteristics of information processing.

Coordination Dynamics, the idea of Scott Kelso is another important example of the new look with respect to understanding the dynamics and plasticity of brain systems (Kelso 2012). This new look considers brain function as a dynamic coordination within and between systems of the brain, that establishes the dynamic processing of input, and the preparation of output. The cortical dynamics may result in both cooperative and competitive interactions within and between systems that result in cognition (Bressler and Kelso 2016). Another example is the use of fMRI and DTI in projects like the Human Connectome, but the time resolution of systems for fMRI and DTI of the brain, are only indirect measurements of electrical changes that are actually occurring on the order of milliseconds—and therefore they are only an approximation of real-time dynamics.

The use of graphic analysis of the correlations of activity in distributed dipole locations, is also a new methodology for differentiation of systems, e.g., Ye et al. (2014). The new look for the recognition in distributed systems, includes the reconstruction of activity in either resting or task based observations, and then the observations into different frequency ranges are filtered to extract regional phase synchrony, calculated within and between regions.

When we introduced the MEG to the British Columbia Downs Syndrome Foundation the funding and enthusiasm we encountered was built around the idea that the MEG could identify the characteristics of specific individual information processing and motor capabilities, for each child who had Down's Syndrome—through the measurements of brain function. Of course the complexity of input and output is an index of the complexity of the system. However, the concept was to individualize training based on brain imaging, and to maximize the capabilities and contributions of each disabled person to the society, and to facilitate the development of themselves—as individuals.

What would this world be like if the new look for brain Imaging is to combine imaging with an increasingly complex computational neuroscience, e.g., the study of brain function in terms of the information processing properties of individual dynamics, and distributed systems, in each individual brain? What are the possible consequences of being able to identify the characteristics of, and potential for, different kinds of behaviour and information processing? What if the dynamic interaction of systems in the brain actually constituted, i.e., defined, "the individual"? What are the distinctions between individual and societal advantages and what if an individual 'consents' to have their brain changed—and then after that changes his or her mind?

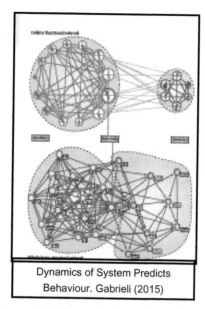

Dynamics of System Predicts
Behaviour. Gabrieli (2015)

There are of course positive and negative consequences of a real understanding of the potential of each individual. At any particular time the consequences can be considered an advantage to the 'society,' as it is defined at that time. What would be the current response of the society if such control were able to stop the killing of people by people—and would that be a good thing? Of course the answer today is no—since the killing of people is current, and has been in the past, implemented primarily for the purpose of control. The question is what are the ultimate consequences of the uses of science, to understand the human brain and whether this will result in a negative or a positive future for the human race.

The technology now is close to being able to identify capability of different system configurations for different types of information, and the degree to which they are unique to different individuals.

What would be the consequences of Recognition Indices for each person? A review, published in the journal Neuron, highlights a number of recent studies showing that brain imaging can help predict an individual's future learning, their future criminality, their health-related behaviours, and their potential response to drug or behavioural treatments. The developing technology may offer opportunities to personalize educational and clinical practices. An example is the studies of Gabrieli and his colleagues (2015) of the Massachusetts Institute of Technology in Cambridge, who describe the predictive power of brain imaging across a variety of different future behaviours, including the infants' later performance in reading, in math and the likelihood of them becoming repeat offenders, of adolescent future drug and alcohol use.

"Presently, we often wait for failure, in school or in mental health, to prompt attempts to help, but by then a lot of harm has occurred" says Gabrieli. "If we can use Neuroimaging to identify individuals at high risk for future failure, we may be able to help those individuals avoid such failure altogether." The authors also point to the clear ethical and societal issues that are raised by studies attempting to predict individuals' behavior. "We will need to make sure that knowledge of future behavior is used to personalize educational and medical practices, and not be used to limit support for individuals at higher risk of failure." For example, rather than simply identifying individuals to be more or less likely to succeed in a program of education, such information could be used to promote differentiated education for those less likely to succeed in the standard education program.

A critical element of the educational system may be the understanding, encouragement and acceptability of diversity of complex systems in the brain, and the recognition of the uncertainty principle in the understanding of those systems, i.e., observation of a system in disturbs the system.

I think the concept of how we understand the world around us is changing, from a focus on central tendency to an understanding of variability. An understanding of variability could be more important than the understanding of central tendency. The importance and function of variability for survival of any system, is beginning to emerge regardless how molar or molecular that system is.

As soon as any system becomes homogenized and variability disappears the system become completely unable to change, and dies. Variability of function in the brain is an index of processing, the analysis of input and the preparation for output all of which is variable and was, and is, instrumental in how the species, any species, survived. The importance of variability is universal and the understanding of it, not just the record of it, is critical in understanding underlying principles of how humans, of how societies, function.

Of course it has always been clear that the homogenization of ideas results in the process of control, this a primary element in politics and religion. If one were to consider the broader implications of this with respect to the nature of our society as a whole, the question is whether survival of a society depends on acceptability of diversity. It has always been clear that the homogenization of ideas results in the process of control. If one were to consider the broader implications of this with respect to the nature of our society as a whole, the question is whether survival of a society depends on acceptability of diversity.

Well, I guess what I suggested at the beginning of this is that the pendulum has begun to swing, from a focus on localized sources in the brain to an understanding of complex, distributed, processing systems.

But what happens when the pendulum begins to swing back again—to an analysis of molecular attributes of the 'system', e.g., the effect of a small localized changes that could make dynamic changes throughout the brain (the Butterfly Effect). The bottom line is that the acceptance of diversity requires a configuration of when the herd mentality is not acceptable—and how is that decided?. Is there a method for increasing plasticity of the brain, that is, increase its capabilities to change even though those changes do not decrease survival of the species—or—to put it differently is there a method of changing the brain to accept diversity?

Therefore the final question is: Is it possible that future technology will result in a change in the Herd Mentality due to an increasing plasticity of the brain? A consequence could be an acceptability of diversity, and be the beginning of the way in which we can "All Live Together".

References

Bressler, S. L., & Kelso, J. A. S. (2016). Coordination dynamics in cognitive neuroscience. *Frontiers in Neuroscience.* http://dx.doi.org/10.3389/fnins.2016.00397 (15 September 2016).

Edelman, G. (1989). *The remembered present, a biological theory of consciousness.* Basic Books.

Gabrieli, J., et al. (2015). Prediction as a humanitarian and pragmatic contribution from human cognitive neuroscience. *Neuron, 85*(1).

Kelso, J. A. S. (2012). Multistability and metastability: Understanding dynamic coordination in the brain. *Philosophical Transactions of the Royal Society B Biological Sciences, 367,* 906–918.

Kievita, R. A., et al. (2016). A watershed model of individual differences in fluid intelligence. *Neuropsychologia, 91,* 186–198.

Nunez, P. (1995). *Neocortical dynamics and human EEG rhythms.* Oxford University.

Rothbart, M. K., & Posner, M. (2015). The developing brain in a multitasking world. *Developmental Reviews, 35,* 42–63.

Skarda, C. A., & Freeman, W. J. (1990). Chaos and the new science of the brain. *Concepts Neuroscience, 1*(2).

Weinberg, H. (1972). The recognition index: A pattern recognition technique suitable for noisy signals. In *Electroencephalography and Clinical Neurophysiology Meetings* (London, England, January, 1972).

Weinberg, H., & Cooper, R. (1972). The recognition index: A pattern recognition technique suitable for noisy signals. *Electroencephalography and Clinical Neurophysiology, 33,* 608–613.

Weinberg, H., Walter, W. G., & Crow, H. J. (1969). Intracerebral events in humans related to real and imaginary stimuli. *Electroencephalography and Clinical Neurophysiology, 27,* 665.

Weinberg, H. W., et al. (1974). Emitted cerebral events. *Electroencephalography and Clinical Neurophysiology, 36,* 449–456.

Weinberg, H., Carson, P., Joly, R., Jantzen, K. J., Cheyne, D., & Vincent, A. (1999). Measurement and monitoring of the effects of work schedule and jet lag on the information processing capacity of individual pilots. *Journal of Aviation Psychology.*

Ye, A. X., Leung, R. C., Schafer, C. B., Taylor, M. J., & Doesburg, S. M. (2014). Atypical resting synchrony in autism spectrum disorder. *Human Brain Mapping, 35,* 6049–6066.

Variability in Life Can Facilitate Learning to Live Together

Nick Stergiou

In this paper I propose that happiness can be achieved by being variable in everything that we do. Being variable results in being adaptable in difficult situations such as dealing with other people that are different to you. By promoting variability in our lives, we can become more contented as we can overcome many obstacles. Striving to be variable in every aspect of our lives can promote a happier and healthier life that can translate to improved tolerance and adaptability in our social interactions. Variability in our lives can thus facilitate learning to live together.

Learning to live together (LLT) can be defined as a process of acquiring knowledge about others and their way of living, in order to co-exist in harmony. United Nations has even developed "courses" to promote LLT since there is a belief that such a process will result in fewer conflicts, an increased tolerance with respect to differences among cultures, and the promotion of peace and world happiness. Therefore, in these courses students are exposed to other cultures, religions and ways of living in order to reduce ignorance.

My proposition to achieve LLT and its associated goals is different. I believe that if we are happier ourselves, we can tolerate others and their differences and can co-exist more effectively. Consider the following. If you are in a very happy state, someone can insult you or may even hit you, but you may actually laugh or dismiss it in your overall state of happiness. However when you are unhappy, you can be irritated much easier and your tolerance level is pretty low regarding anything that you don't like or want.

Therefore, I propose that if we want to promote LLT, we may be able to achieve significant results through the promotion of personal happiness. This creates a critical question: how can I achieve personal happiness? This is a question that has been asked by almost everybody through the centuries and has created volumes of

N. Stergiou (✉)

Division of Biomechanics and Research Development, University of Nebraska, Omaha, USA

e-mail: nstergiou@unomaha.edu

© Springer International Publishing AG, part of Springer Nature 2019

J. A. S. Kelso (ed.), *Learning To Live Together: Promoting Social Harmony*,

https://doi.org/10.1007/978-3-319-90659-1_8

books and manuscripts. However my approach is unique and quite mechanistic. I propose that happiness could be achieved by being variable in everything that we do. Being variable results in being adaptable in difficult situations, including when dealing with people that are different to us.

To allow for a better understanding of this proposition, as an example I will use movement, which is the area within which I conduct my research. When we perform the same task multiple times, we can easily observe that we never execute it the exact same way. This is obvious even with elite performers such as athletes and musicians. When it comes down to simple everyday walking every one of us is an elite performer, but as we observe our steps behind us on the sand we can clearly see that they are never identical. These natural fluctuations in motor performance define the presence of variability which is ubiquitous in all biological systems. In other words, variability is a fact of life (Harrison and Stergiou 2015; Stergiou and Decker 2011).

For years variability was considered as an indicator of undesirable noise in the control system (Newell and Corcos 1993; Schmidt and Lee 2005). This position was likely due to the common use of linear statistical measures (e.g. standard deviation) to study variability (Stergiou et al. 2006). Such measures contain no information about how the motor system responds to change over time. Practically, the linear measures are measures of centrality and thus provide a description of the amount or magnitude of the variability around that central point. This is accomplished by quantifying the magnitude of variation in a set of values independent of their order in the distribution. From this perspective, clinicians and scientists believe that the mean is the "gold standard" of healthy behavior. Any deviation from this gold standard is error, or undesirable behavior, or the result of instability. However, recent literature from several disciplines including brain function and disease dynamics have shown that many apparently "noisy" phenomena are the result of nonlinear interactions and have deterministic origins (Amato 1992; Buchman et al.2001; Cavanaugh et al. 2010; Garfinkel et al. 1992; Goldstein et al. 1998; Orsucci 2006; Slutzky et al. 2001; Toweill and Goldstein 1998; Wagner et al. 1996). In movement, the study of variability was massively enhanced by instrumentation that has allowed scientists to put movement literally under the microscope. Cameras that can capture movement in hundreds of pictures per second (our eye can only see 12–15 pictures per second) and computers that can store and process with amazing speeds such an abundance of data, has allowed us to observe movement with detail that parallels what Koch and Fleming were able to see when they had blood under their microscopes.

As a result, it was found that stride-to-stride variations in healthy walking exhibit nonlinear and fractal-like fluctuations extending over hundreds of steps. The classic definition of a fractal, first described by Mandelbrot (Mandelbrot 1977), is a geometric object with "self-similarity" over multiple measurement scales (Stergiou 2016). The outputs of the locomotor system measured over time exhibit such fractal properties (Delignieres and Torre 2009), demonstrating power-law scaling such that the smaller the frequency of oscillation (f) of these signals, the larger their amplitude (amplitude squared is power) (Harrison and Stergiou 2015; Stergiou 2016). This power-law relation can be expressed as 1/f, and is referred to as pink noise, where oscillations appear self-similar when observed over seconds, minutes, hours, or days. In terms of a distribution, this means that when we naturally walk, the variations of our strides are not normally distributed but instead we have a few big strides, many medium size strides, and a huge number of small size strides.

Importantly, such nonlinear and fractal-like 1/f distributions exist everywhere around us (e.g. in trees, lightning, cloud formation) and inside us (e.g. in our airways, intestine folds, blood vessels). Furthermore, we also have an affinity for such patterns. A work of art is pleasing if is neither too regular/predictable, nor packs too many surprises. Such patterns are ubiquitous in human performance that we observe in music, art, elementary motor tasks, cognitive tasks, etc.

These observations have allowed the development of new theoretical frameworks to study movement variability. Thus, it has been proposed that the natural fluctuations that are present in normal motor tasks (e.g. stride-to-stride fluctuations in normal walking) are characterized by an appropriate or optimal state of variability (Harrison and Stergiou 2015; Stergiou et al. 2006; Stergiou and Decker 2011). Optimal variability is associated with nonlinear and fractal-like 1/f distributions. This physiologic complexity enables an organism to function and adapt to the demands of everyday life (Harrison and Stergiou 2015; Stergiou et al. 2006; Stergiou and Decker 2011). This physiological complexity is recognized as an inherent attribute of healthy biological systems, whereas the loss of complexity, for example with aging and disease, is thought to reduce the adaptive capabilities of the individual. A loss of complexity results in an overly constrained, periodic and rigid system, or an overly random, noisy, incoherent system.

There are compelling findings in both animal and human studies that suggest that the complexity of locomotor patterns provide a rich source of information that could be relevant to the diagnosis and management of a variety of diseases that affect an aging population. Our previous research has shown that highly active older adults exhibit more complex patterns of locomotor activity than less active older adults, despite the absence of differences between these groups in standard measures of variability of their step counts (Cavanaugh et al. 2010). Hu and colleagues have recently shown that older adults and dementia patients have disrupted fractal activity patterns (Hu et al. 2009) and that the degree of disruption is positively related to the burden of amyloid plaques—a marker of Alzheimer's disease severity (Hu et al. 2013). They also found that fractal scaling in activity fluctuations is unrelated to the average level of activity as assessed within and between subjects (Hu et al. 2004). A study of primates suggests that a loss of complexity in locomotor behaviour that is associated with illness and aging, reduces the efficiency with which an animal is able to cope with heterogeneity in its natural environment (Macintosh et al. 2011). Japanese quail became less periodic and more complex in their locomotor behaviour when they were stimulated to explore, without there being commensurate changes in the percentage of total time spent walking, or in the average duration of the walking events (Kembro et al. 2009). Additionally, fractal scaling has been observed in the locomotor activity of young, healthy small mammals, a feature that is less evident in aged animals (Anteneodo and Chialvo 2009).

In summary, an optimal level of variability enables us to interact adaptively and safely to a continuously changing environment, where often our movements must be adjusted in a matter of milliseconds. A large body of research exists that demonstrates natural variability in healthy gait (along with variability in other, healthy biological signals e.g. heart rate), and a loss of this variability in ageing and injury, as well as in a variety of neurodegenerative and physiological disorders. In 1944, Erwin Schrodinger said something similar in his book "What is Life". Specifically he stated that "Life is an aperiodic crystal, it is not random, but also is not periodic, it is something in between." From my perspective variability may be the spice of life.

Now let us return to LLT and my proposition. As a reminder, I have proposed that happiness could be achieved by being variable in everything that we do. Being variable results in being adaptable in difficult situations such as dealing with other people that are different to you. By promoting variability in our lives, we can become much happier as we can overcome many obstacles. This approach also keeps as healthier which further enhances happiness.

Recommendations

How can we promote variability in our lives? In many different ways. For example, when we drive to work in the morning we can purposely select different pathways. At the same time we can also use fractal-like 1/f distributions. For example, when we eat we can be variable in terms of the types of foods and their calories. We can eat a few pieces of food that have many calories (i.e. chocolate), more pieces of food that have less calories (i.e. meat), and a huge number of food that have a very small amount of calories (i.e. vegetables). Isn't actually this type of a distribution what is recommended by almost every diet that promotes health?

In conclusion, striving to be variable in every aspect of our lives can promote a much happier and healthier life that can easily translate in improved tolerance and adaptability in our social interactions. Variability in our lives can thus facilitate learning to live together.

References

Amato, I. (1992). Chaos breaks out at NIH, but order may come of it. *Science (New York, N.Y.)*, *256*(5065), 1763–1764.

Anteneodo, C., & Chialvo, D. R. (2009). Unraveling the fluctuations of animal motor activity. *Chaos (Woodbury, N.Y.)*, *19*(3), 033123. https://doi.org/10.1063/1.3211189.

Buchman, T. G., Cobb, P. J., Lapedes, A. S., & Kepler, T. B. (2001). Complex systems analysis: A tool for shock research. *Shock*, *16*(4), 248–251.

Cavanaugh, J. T., Kochi, N., & Stergiou, N. (2010). Nonlinear analysis of ambulatory activity patterns in community-dwelling older adults. *The Journals of Gerontology. Series A: Biological Sciences and Medical Sciences, 65*(2), 197–203.

Delignieres, D., & Torre, K. (2009). Fractal dynamics of human gait: A reassessment of the 1996 data of hausdorff et al. *Journal of Applied Physiology (Bethesda, Md.: 1985), 106*(4), 1272–1279. https://doi.org/10.1152/japplphysiol.90757.2008.

Garfinkel, A., Spano, M. L., Ditto, W. L., & Weiss, J. N. (1992). Controlling cardiac chaos. *Science (New York, N.Y.), 257*(5074), 1230–1235.

Goldstein, B., Toweill, D., Lai, S., Sonnenthal, K., & Kimberly, B. (1998). Uncoupling of the autonomic and cardiovascular systems in acute brain injury. *The American Journal of Physiology, 275*(4 Pt 2), R1287–92.

Harrison, S. J., & Stergiou, N. (2015). Complex adaptive behavior and dexterous action. *Nonlinear Dynamics, Psychology, and Life Sciences, 19*(4), 345–394.

Hu, K., Ivanov, P. C., Chen, Z., Hilton, M. F., Stanley, H. E., & Shea, S. A. (2004). Non-random fluctuations and multi-scale dynamics regulation of human activity. *Physica A: Statistical Mechanics and its Applications, 337*(1), 307–318.

Hu, K., Van Someren, E. J., Shea, S. A., & Scheer, F. A. (2009). Reduction of scale invariance of activity fluctuations with aging and alzheimer's disease: Involvement of the circadian pacemaker. *Proceedings of the National Academy of Sciences of the United States of America, 106*(8), 2490–2494. https://doi.org/10.1073/pnas.0806087106.

Hu, K., Harper, D. G., Shea, S. A., Stopa, E. G., & Scheer, F. A. (2013). Noninvasive fractal biomarker of clock neurotransmitter disturbance in humans with dementia. *Scientific Reports, 3*, 2229. https://doi.org/10.1038/srep02229.

Kembro, J. M., Perillo, M. A., Pury, P. A., Satterlee, D. G., & Marin, R. H. (2009). Fractal analysis of the ambulation pattern of japanese quail. *British Poultry Science, 50*(2), 161–170.

Macintosh, A. J., Alados, C. L., & Huffman, M. A. (2011). Fractal analysis of behaviour in a wild primate: Behavioural complexity in health and disease. *Journal of the Royal Society, Interface/ the Royal Society, 8*(63), 1497–1509. https://doi.org/10.1098/rsif.2011.0049.

Mandelbrot, B. B. (1977). *The fractal geometry of nature.* New York: W.H. Freeman and Company.

Newell, K. M., & Corcos, D. M. (Eds.). (1993). *Variability and motor control.* Champaign IL: Human Kinetics Publishers.

Orsucci, F. F. (2006). The paradigm of complexity in clinical neurocognitive science. *The Neuroscientist: A Review Journal Bringing Neurobiology, Neurology and Psychiatry, 12*(5), 390–397. doi:12/5/390 [pii].

Schmidt, R. A., & Lee, T. D. (2005). *Motor control and learning: A behavioral emphasis.* Champagne, IL: Human Kinetics.

Slutzky, M. W., Cvitanovic, P., & Mogul, D. J. (2001). Deterministic chaos and noise in three in vitro hippocampal models of epilepsy. *Annals of Biomedical Engineering, 29*(7), 607–618.

Stergiou, N. (Ed.). (2016). *Nonlinear analysis for human movement variability* CRC Press.

Stergiou, N., & Decker, L. M. (2011). Human movement variability, nonlinear dynamics, and pathology: Is there a connection? *Human Movement Science, 30*(5), 869–888.

Stergiou, N., Harbourne, R., & Cavanaugh, J. (2006). Optimal movement variability: A new theoretical perspective for neurologic physical therapy. *Journal of Neurologic Physical Therapy, 30*(3), 120–129.

Toweill, D. L., & Goldstein, B. (1998). Linear and nonlinear dynamics and the pathophysiology of shock. *New Horizons (Baltimore, Md.), 6*(2), 155–168.

Wagner, C. D., Nafz, B., & Persson, P. B. (1996). Chaos in blood pressure control. *Cardiovascular Research, 31*(3), 380–387. doi:0008636396000077 [pii].

Bridging the Gap

Jane L. Buck

The hotly contested 2016 presidential campaign in the United States revealed socio-political fissures that have been widening for decades. A primary contributor to those divides is an enormous and increasing income disparity, one which disproportionately affects persons of color. I take the position that civil harmony will be difficult, if not impossible, to achieve without good-faith efforts to reduce the income gap. I offer two arguably utopian policy proposals, both of which will admittedly be difficult to implement, but which could allow us to make progress toward the goal of a more equitable distribution of income. First, cities should make public transportation free or heavily subsidized. This would open opportunities to workers to seek employment in places previously inaccessible. Second, large corporations that reduce their workforce for whatever reason should be required to find work for their laid-off employees. They should retrain them if necessary, and, if new jobs require relocation, the company should bear the cost. The implementation of these proposals will probably require extensive public-private co-operation as well as creative approaches to financing.

Over half a century ago, Allport (1954) hypothesized that increased contact between groups would facilitate understanding and reduce conflict and bias. The degree to which contact would have the desired effect is contingent upon a variety of factors: the duration of the contact, the relative status of the groups involved, the milieu in which the contact takes place, and the purpose of the contact, *inter alia*.

My personal experiences lend anecdotal support to aspects of Allport's contact hypothesis. I spent the first seven years of my life in a rural corner of southeastern Pennsylvania overwhelmingly populated by the descendants of 18th Century German immigrants. A few of my first-grade classmates spoke only Pennsylvania German, despite their families' long history in the United States, and most of my maternal grandfather's veterinary clients spoke no English. They accused my

J. L. Buck (✉)
Department of Psychology, Delaware State University, Dover, DE, USA
e-mail: janebuckphd@gmail.com

© Springer International Publishing AG, part of Springer Nature 2019
J. A. S. Kelso (ed.), *Learning To Live Together: Promoting Social Harmony*,
https://doi.org/10.1007/978-3-319-90659-1_9

mother of marrying a foreigner, because my father, also the descendent of German settlers, was born in the county seat 15 miles away, where his father's closest friend was African-American.

When I turned seven, we moved to a steel-and-coal town in eastern Ohio, where I encountered for the first time children of many ethnicities. Our classrooms were fully integrated, and I recall no overt racial or ethnic bias. I was, however, painfully aware that my German accent set me apart, and I quickly learned to eliminate it.

A few years later we moved to Delaware, a former slave state that had fought on the side of the North in America's Civil War. It took me some time to understand why I felt so uncomfortable in my new surroundings. There were no "colored" children in my school. During the next several years my only contact with persons of color was through my father's professional contacts as a Boy Scout executive. He had an African American colleague, who was an occasional guest in our home, and a part of their responsibilities was to attempt to integrate the local organization. At the time there were separate camping facilities for black and white Scouts. I recall one particularly painful incident when my father was called in the middle of the night to go to the whites-only camp to pick up a black Scout who had been sent there in error and transport him to the blacks-only camp.

When I matriculated at the University of Delaware five years later, my parents and I were guests at a luncheon for new students and their families. My father pointed to a black waiter and informed us that the man had a Ph.D. in chemistry. Waiting tables was the only job he could get in the strictly segregated border state. Of course I had no black classmates. My African American contemporaries seeking post-secondary education had two options: leave the state or attend Delaware State College, at that time an underfunded, struggling, unaccredited, and blacks-only institution. About a year later ten Delaware State College students filed a successful suit demanding the right to attend the University. The following year a small group of black students were admitted.

Delaware's public elementary and high schools remained completely segregated under the "separate but equal" doctrine until 1954, when the United States Supreme Court ruled in a historic and unanimous decision in *Brown v. Board of Education* that separate facilities are inherently unequal. What is often forgotten is the fact that the case consolidated suits from four states, including Delaware. The court chose not to prescribe a remedy, given the complex and varying circumstances of the segregated states and instead ordered that they desegregate "with all deliberate speed."

The assassination of Dr. Martin Luther King, Jr. in April 1968 precipitated months of racial unrest. The National Guard occupied Wilmington, Delaware, for seven months, and Delaware State College was forced to close as students protested and the Guard controlled the campus. At the time, I was completing my doctorate in behavioral sciences at the University of Delaware, while employed as a research associate with a federally funded research laboratory with an office on campus. The University, despite being formally desegregated, had only a relative handful of black students. The faculty, with few exceptions, was white.

Just across the state line in Pennsylvania, students at Cheney State College, which claims to be the oldest historically black college in the country, continued protests that had begun earlier over complaints regarding inadequate facilities and lack of student involvement in the governance of the institution. The situation escalated to the point that the College's president was forced to resign. He found a new home on the faculty of the University of Delaware and raised the proportion of black faculty by a substantial amount. We saw one another on a fairly regular basis. In the spring of 1969, after I had left the regional laboratory, he urged me to apply for a faculty position at Delaware State. Following his advice, I applied for and was hired in a half-time position starting in the fall.

The student demographics of the College and the University were essentially mirror images. There were very few white students at Delaware State, although the faculty was approximately evenly divided. Almost two decades after the court ordered that the public higher educational institutions integrate, the disparity in financial support was painfully obvious in everything from the physical environment to the number of volumes in the libraries, the number of faculty members with doctoral degrees, and faculty salaries. The University, with a sizable endowment, was superior in almost every respect to the College, which was almost completely reliant on the dubious largesse of the state.

It is one thing to bemoan the circumstances of others while viewing them from a privileged distance and another to experience them firsthand. Reading that the University's library has nine times as many volumes as the College's is quite different from attempting to create a reading list for one's students with readily available titles. It is one thing to be aware in an abstract way that young black men are arrested and incarcerated at rates far exceeding those of young white men, quite another to witness an excellent student, a member of a service fraternity, brought in chains before a judge and sentenced to prison for driving with an expired driver's license.

Progress towards the goal of integration at all levels was characterized more by deliberation than by speed. As *The New York Times* reported (1977), 23 years after the Supreme Court decision, Delaware was still engaged in a struggle to end segregation in Wilmington, where the public school population was 85% black. A Federal court ordered that the city's approximately 10,000 black students be integrated with the suburbs' 62,000 mostly white students through cross-district busing. The remedy was almost universally unpopular.

Forbes (1997) reviewed a study by Parsons (1985) that measured the racial attitudes of black and white parents and students in New Castle County, where Wilmington is located. Data were gathered annually starting in 1978, prior to the implementation of busing, until 1981. The results were disappointing to those who believed in a simplistic version of the contact hypothesis that mere exposure is sufficient to improve race relations. Not only did racial attitudes not improve in any significant manner, they worsened somewhat over time, although not in a statistically significant way. Forbes pointed out that the study had many methodological flaws, including a short time period, nonrandom sampling, and a return rate for questionnaires of only two percent. Added to these was the fact that opposition to desegregation, including further litigation, continued during the period covered. It is not unreasonable to speculate that any potentially positive benefits of increased contact might have been nullified by its coercive nature.

At the post-secondary level both the University of Delaware and Delaware State College, now known as Delaware State University, struggled and continue to struggle to integrate. During my almost three-decade-long tenure at Delaware State, racial demographics fluctuated widely, by my estimate, from approximately 98% to approximately 60% black. According to the National Center for Education Statistics (2017), the University of Delaware, as recently as the fall of 2015, had a black enrollment of just under 6% in a state whose black population, according to U.S. Census data (2010), constitutes 22% of the total. Delaware State's black enrollment was 70%. The reasons for this intransigent de facto segregation include cost and admission standards. The University of Delaware is both more expensive and more selective than Delaware State. Additionally, many first-generation African American students express a desire to attend a historically black institution where they feel more comfortable.

Forbes (1997) points out that racial attitudes are more likely to improve when groups interact over long periods and in circumstances of relative equal status. To this end, I propose two potential solutions, the first of which has been implemented in the small city, Newark, Delaware, that is the home of the University of Delaware. Free or heavily subsidized public transportation could provide the means for those living in poverty to obtain employment in otherwise inaccessible areas, leading to increased interaction among groups, which in turn might improve relations.

The City of Newark and the University of Delaware, with subsidies from the State of Delaware, provide free bus transportation within the city limits. Admittedly, this model is neither perfect nor necessarily adaptable to every other municipality, but it is a place to start. One drawback is a schedule that is quite limited owing to a lack of resources sufficient to provide more comprehensive coverage. For example, one needing a round trip ride between points A and B might find that the trip from A to B is ten minutes long, while the return trip is an hour or more. Nevertheless, the service provides mobility to those who cannot afford other means of transportation.

Funding for the service is provided by the City, the University, and in most years, the State. The City Manager (Houck 2016) wrote in a memorandum to the City's mayor and council that the Delaware Transit Corporation has provided operating funds to the service (known as UNICITY) since 1983. The University has supplied personnel and supervision and assumed operational control. The City has assumed the costs of bus maintenance, fuel, and insurance premiums.

Clearly, Newark's model requires the co-operation and co-ordination of entities that might well be unavailable in many, if not most, cases. The involvement of private business could substitute for that of the University. The challenge would be to convince business owners that access to a larger and more diverse workforce is to their advantage.

My second policy proposal is admittedly more radical and one for which I have no direct experience. Large businesses that reduce their workforces in the service of unrestrained profit-seeking should be required to find new employment with comparable wages for their laid-off employees. Should that require retraining or relocation, as it probably would, the company should be required to bear the total cost or be subject to fines sufficient to allow the government to do so. Just as most large businesses set aside funds for the replacement of aging equipment and for pensions, they should create contingency funds to meet the needs of labor subject to layoff. Predatory capitalism that exploits its workers exacts a societal and economic price in the form of poverty and its attendant ills: increased rates of crime, drug and alcohol abuse, infant and maternal mortality, and racial tension, to name the most obvious. Obscenely overcompensated managers should not be allowed to shift the costs of their labor practices to the larger society.

These proposals might legitimately be seen as utopian and, even if implemented, would not be panaceas. But they might make a modest beginning in the effort to reduce racial and ethnic conflict.

References

Allport, G. W. (1954). *The nature of prejudice*. Cambridge: Addison-Wesley Publishing Co.

Forbes, H. D. (1997). *Ethnic conflict: Commerce, culture, and the contact hypothesis*. New Haven and London: Yale University Press.

Franklin, B. A. (1977). Wilmington desegregation fight appears heading for a showdown. *The New York Times*, 10.

Houck, C. S. (2016). (Personal communication).

National Center for Education Statistics. (2017). Retrieved from https://nces.ed.gov/globallocator.

Parsons, M. A. (1985). Parents = and students = attitude changes related to school desegregation in New Castle County, Delaware, in *Metropolitan Desegregation*, (Ed.) R. I. Green, New York, Plenum.

United States Census. (2010). Retrieved from https://www.census.gov/population/race/data/cen2010.html.

Walls and Borders and Strangers on the Shore: On Learning to Live Together from the Perspective of the Science of Coordination and the Complementary Nature

J. A. Scott Kelso

Man tries to make for himself in the fashion that suits him best a simplified and intelligible picture of the world; he then tries to some extent to substitute this cosmos of his for the world of experience, and thus to overcome it. This is what the painter, the poet, the speculative philosopher, and the natural scientist do, each in his own fashion. Each makes this cosmos and its construction the pivot of his emotional life, in order to find in this way the peace and security which he cannot find in the narrow whirlpool of personal experience.

A. Einstein

Overview. To learn to live together requires a world mind-shift. In this paper I will attempt to articulate what this mind-shift involves and how it is constituted. As remarked by previous Olympians of the Mind (the solution to learning to live together rests ultimately not on science or technology or economics or politics, but on human decency and compassion. Humanity must realize that our collective fates are intertwined both in terms of uniqueness and interdependence. Regardless of sex, race, religion, economic opportunities, individual passions or ambitions, we must somehow weave a "we" to see native and stranger on the same footing. We will not learn to live together until we face this collective reality. The science of coordination (Coordination Dynamics) and the philosophy that arises from and underlies this science (The Complementary Nature) offer a way to change the world to one we all need, one where together we can live in a truly relational way that is without prejudice and goes beyond simple tolerance of the other. Some of the key empirically-based concepts I will discuss are *synergy*—the "working together" aspect as a self-organized entity (in the sense of the physics of open, nonequilibrium systems) and as the significant unit of biological

J. A. S. Kelso (✉)
The Human Brain & Behavior Laboratory, Center for Complex Systems & Brain Sciences, Florida Atlantic University, Boca Raton, FL, USA

J. A. S. Kelso
The Intelligent Systems Research Centre, Ulster University (Magee Campus), Derry ∼ Londonderry, Northern Ireland

J. A. S. Kelso (ed.), *Learning To Live Together: Promoting Social Harmony*,
https://doi.org/10.1007/978-3-319-90659-1_10

coordination (in the sense of synergistic selection); *learning*—the modification of pre-existing biases and dispositions; the *nature of change*—a process unpredictably sudden and abrupt or slow and tortuous depending on identifiable competitive or cooperative mechanisms; *agency*—a fundamentally relational and dynamic attribute not isolated in the individual mind; and finally, the *metastable dynamics* of the human brain—how the tendencies for the parts of the brain to integrate co-exist with tendencies for individual autonomy and segregation. I will present new experimental evidence which demonstrates that a critical level of diversity separates these two idealized régimes. Whereas bistability is the basis for polarized either/or thinking and phase transitions, which allow one to switch from one polar extreme to the other, the in-between metastable régime—which contains neither stable nor unstable states (no states at all in fact)—gives rise to a far more fluid, complementary mode of operation (hence, The Complementary Nature) in which it is possible for apparent contrarieties (e.g., integration $\sim$ segregation, unity $\sim$ diversity, individual $\sim$ collective, self $\sim$ other, cooperation $\sim$ competition, chance $\sim$ choice, boundary $\sim$ domain, etc., etc.) to coexist in the mind at the same time. The political, ethical and educational consequences of the metastable brain $\sim$ mind that sees contrarieties as complementary are many, including a fundamentally "new" triadic logic not of the excluded (after Aristotle) but of the *included* middle, signified by the tilde or squiggle ($\sim$) symbol. The metastable brain $\sim$ mind, if we can tap into it, signals the end of dualism, the grand "either/or," and "the perpetual contradiction of opposites" that is at the very core of religious and ideological conflict throughout history.

Preamble. As I sit down to write my thoughts on "Learning to Live Together", Hurricane Irma is heading toward the South of Florida, where I live, at 185 miles per/h. People are afraid, and rightly so, but also in part because a major tragedy occurred in Texas recently when Hurricane Harvey struck. Prior context matters. Fear brings out the worst in people. When resources are scarce and time is short, people tend to behave badly. One only has to watch them when they are queuing up to fill their cars with petrol or 'gas' (in American English). Their movements and gestures are jittery and the language is coarse. The slightest change can trigger a strong reaction. Ordinary discourse goes out the window, replaced by a snarl here and a 'fuck you' there. If you don't have a fueled means of transport to escape the storm, you and your kin are trapped, man. Gas is running out and competition is fierce. There's a reason why the old Darwinian slogan of "survival of the fittest" has lasted over the years. It's me against you and it's ugly.

Yet, you only have to observe people's behavior after the storm in Texas to see plenty of evidence for their good side. The giving, the sharing, the brave, sometimes heroic deeds to save others in distress—it warms the heart to see it. High levels of empathy in the TV viewing audience are invoked followed by material, philanthropic giving. Strange things us humans. Fiercely competitive on the one hand, cooperative and altruistic on the other. How are we to understand the nature of man?

These passing thoughts are not unrelated to our topic of learning to live together. The availability of resources does matter. Our evolutionary origins and history do matter. How we were brought up does matter. Environment and education do matter. Inequality is an issue. Etcetera, etcetera, etcetera as the King of Siam used to say. But I don't think any of these factors are the true heart of the problem. In my view, to learn to live together requires a world mindshift. My aim in this paper is a small attempt to articulate what the mindshift might involve, how it is constituted. Importantly, it's not only about the mindshift itself, but how it fits into a world that is heavily polarized on many levels. I should say also that the world I am talking about is most likely limited to those in so-called democracies who at least think they can do something about it—though in actuality their voice is often muted and smothered. In every country, in every continent, the majority of people cannot do a damned thing. In my view they will stay silent until they think and act in a certain way and have the will to do something about it.

The sculpture shown in Fig. 1 is called "Hands across the Divide" and stands at the end of the Craigavon bridge in Derry, in the northern part of Ireland—the place where I am from–and of course is symbolic of the hope for an end to conflict and division. In this, the sculpture resonates with the 8th and 9th Olympiads of the Mind (and indeed earlier Olympiads—all organized by an extraordinary human being, Dr. Epimenides Haidemenakis) which aim to cultivate greater understanding, tolerance and unity among human beings worldwide.

What can science and in particular brain and behavioral research do to end conflict and promote global harmony—to help us learn to live together? On an earlier occasion[1] my position was that despite all the successes of contemporary neuroscience in alleviating the many neuropsychiatric and neurological diseases that afflict us—even removing the stigma of mental illness and epilepsy—not much has really changed. Looking around the world today it is difficult to avoid the conclusion that we human beings are the way we are. Wars, poverty, violence, fear, greed, etc. permeate modern life just as they have for centuries. All our knowledge of the brain, I said then, and all the marvelous technological developments that have helped produce this knowledge, have not led to much wisdom or deeper under-standing of ourselves. To turn scientific knowledge into wisdom, I argued, seems to involve an alchemy that has escaped us....

However, I believe there is light at the end of the tunnel. There are at least five reasons for hope that I will discuss. One is that the 'new science of coordination' called *coordination dynamics* promotes the centrality of *synergy* as the fundamental unit of life. Synergy is from the Greek *sunergia* and means "working together". The significance of synergy is that it places emphasis on the effectiveness of cooperative interaction between two or more agents—related or not—and is the key to our survival and to learning to live together. Two is that the basis for how synergies arise, persist and change in natural systems (including us humans) is known to rest

---

[1]Kelso (2010) Coordination and The Complementary Nature. Nour Foundation, New York. https://www.youtube.com/watch?v=FHd5FLwTspk.

Fig. 1 *Reconciliation/Hands across the Divide* by the Sculptor, Maurice Harron who like the author was born and grew up in Derry, Northern Ireland. The sculpture is situated in Carlisle Square in Derry ~ Londonderry overlooking the Craigavon Bridge which spans the River Foyle. As a result of 'the troubles', the bridge came to largely separate Protestant and Catholic communities. Bridges have a dual function: they can unite or divide

on the joint action of two fundamental forces of nature: evolution and self-organization. Three is that synergies may be expressed in a common theoretical language, that of informationally meaningful, (predominantly) bidirectionally coupled, nonlinear dynamical systems (Coordination Dynamics). Four is that the empirical and mathematical study of the latter over the last 30 years in laboratories and research centers around the world have revealed a feature called *metastability* that, among other attributes, has been hailed as a *new principle of brain function*. Fifth, is that this new principle leads directly to a mindset that signals the end of polarization, here considered to be the root cause of strife and conflict. I will touch on all these aspects in what follows.

Coordination Dynamics: The new science of coordination. Many moons ago, my colleagues, students and I set out to understand (if not solve) the problem of coordination in living things. Our work was inspired by the eminent theoretical biologist Howard Pattee (1976) who referred to the problem of coordination as crucial to understanding the physical basis of life. We chose movement, the animated, living movement of human beings as the test field, in part because of a childhood love for sports and the performing arts, and in part because like gravity, most people take their innate capacity to move for granted. Like most scientists at the time, and even more nowadays, we proceeded in classical reductionist fashion.

A first step was to identify *the significant units of biological coordination* and their key properties. This is not a trivial problem nor can it be assumed a priori: animate movement is not made up merely of a list of component parts such as molecules, muscles, neurons and brains, but rather has to do with how these many parts *relate* to each other. Let me dispel any confusion between units *in* and units *of*. The former analyzes units as if they were a piece in a puzzle or an ingredient in a cake. A pendulum, for example, consists of a number of components that can be thought of as the units *in* a pendulum system. But it is *the relations among the components* that define the *function* of the pendulum system. With some notable exceptions (e.g. Noble 2008) biology classifies its units—genes, enzymes, proteins, cells, etc. in terms of their anatomy. The units we were after are units of function[2] which go beyond the particular 'parts list' of components of which they are constituted.

What we found is that the significant units of coordination (and, we think, of life itself) are *functional* synergies or *coordinative structures*—ensembles of interacting neurons, muscles and joints temporarily assembled to accomplish a task or fulfill a function. "Synergies of meaningful movement" (to use the philosopher-biologist Maxine Sheets-Johnstone's coinage) have been hypothesized as important for motor control for over 100 years but until our research in the late 70's and early 80's the evidence was anecdotal or restricted to so-called 'pre-wired' rhythmical activities such as locomotion and respiration. Much work has been done since, of course, and books written (e.g. Kelso 1995; Latash 2008; Sheets-Johnstone 1999/ 2011). So why are synergies preferred over other candidates such as currently popular circuits and networks? Only synergies embrace variability in structure and function (see also Stergiou, this volume). Only synergies handle the fact that many different components can produce the same function (biological degeneracy) and that the same components may be assembled to produce multiple functions (pluripotentiality). Synergies or coordinative structures are not restricted to muscles; they have been identified at many scales from the cellular and neural, to the cognitive and social (e.g. Oullier et al. 2008).

The deeper reasons for synergies as the basic units of biological organization are, as mentioned above, that they are the result of two elemental forces, evolution and self-organization. When cooperation occurs between two or more entities and that cooperation proves to be functionally advantageous, *synergistic selection* is deemed to occur. According to the latter, cooperating groups may gain an advantage in terms of survival and reproduction compared to groups of non-cooperating individuals. This effect is under appreciated though it occurs at all levels of biological organization (see Maynard-Smith and Szathmary 1995; Corning 2010). The other force, unknown to Darwin and mostly ignored by evolutionary biology is *self-organization*—the discovery of emergent cooperative phenomena in natural systems. Of significance here is that the most fundamental form of self-organization in systems that are open to exchanges of energy and matter—as the physicist Hermann Haken has shown comprehensively in his pioneering research on lasers—are nonequilibrium phase transitions (Haken 1977/1983). And of even more significance for us is that all the predicted features of nonequilibrium phase transitions including enhancement of fluctuations and critical slowing down have been demonstrated in coordinated movement and the human brain (Kelso et al. 2013 for review). For matters of movement and mind, self-organizing principles are

---

[2]Notwithstanding the fact that what we call structures, like bones and such, are really slow, relatively long lasting functions. The level and timescale of analysis constrain the terminology. Here, no dichotomy is brokered between structure and function.

expressed in terms of informationally coupled dynamical systems *aka* Coordination Dynamics, Kelso 2009).

A key concept of self-organizing coordination dynamics is the so-called *order parameter* or *collective variable*, a term that expresses cooperative behavior in open, nonequilibrium systems with many degrees of freedom (Haken 1983). It turns out that order parameters (OPs) are important for understanding *any* kind of coordination, from the brain to players in teams, from ballet dancers to championship rowers, because they constitute the *content* of the underlying dynamics (Fuchs and Kelso 2017). Not only are OPs expressions of emergent patterns among interacting components and processes, they in turn modify the very components whose interactions create them. This confluence of top-down and bottom up processes results in *circular or reciprocal causality*, an essential concept in Coordination Dynamics (see Fig. 2).

Unlike the laws of motion of physical bodies, *laws of coordination are expressed as the flow of coordination states produced by functional synergies or coordinative structures.* The latter span many different kinds of things and participate in many processes and events at many scales. In their most elementary form, coordination laws are governed by *symmetry* (and symmetry breaking) and arise from *nonlinear coupling* among the very components, processes and events that constitute the coordinative structure on a given level of description. An example is the well-known extended HKB equation of coordinated movement:

$$\dot{\phi} = \delta\omega - a\sin\phi - 2b\sin 2\phi + \sqrt{Q}\xi_t \tag{1}$$

Fig. 2 *The circular or reciprocal causality of self-organizing coordination dynamics.* Collective coordination patterns characteristic of a functional synergy or coordinative structure on a given level of description arise from the interaction among variable subsystems and processes (upward causation) yet reciprocally constrain the behavior of these coordinating elements (downward causation)

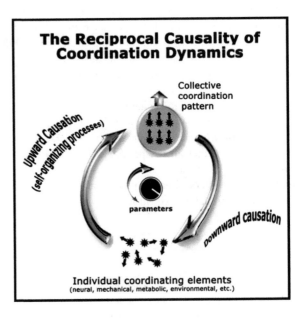

where ϕ is the Order Parameter, in this case the relative phase between two interacting components, the dot above ϕ standing for the derivative with respect to time, a and b are coupling parameters, $\sqrt{Q}\xi_t$ is a (delta-correlated) noise term of strength Q, and $\delta\omega$ is a symmetry breaking term expressing the fact that each coordinating element possesses its own intrinsic behavior. Akin to the Schrödinger equation which describes how the quantum state of a system evolves over time, Eq. (1) specifies how the coordination states of a system evolve over time. Figure 3 shows the layout of attractors of this elementary coordination law.

When $\dot{\phi}$ reaches zero (flow line becoming white), the system ceases to change and fixed point behavior is observed. Note that the fixed points here refer to *emergent coordinative states* produced by the nonlinearly coupled elements that constitute the synergy or coordinative structure. Stable and unstable fixed points at the intersection of the flow lines with the isoplane $\dot{\phi}=0$ are represented as filled and open circles respectively. To illustrate the different régimes of the system, three representative lines labeled 1 to 3 fix $\delta\omega$ at increasing values. Following the flow line 1 from left to right, two stable fixed points (filled circles) and two unstable fixed points (open circles) exist. This flow belongs to the multistable (here *bistable*) régime. Following line 2 from left to right, one pair of stable and unstable fixed points is met on the left, but notice the complete disappearance of fixed point behavior on the right side of the figure. That is, a qualitative change (*bifurcation; phase transition*) has occurred due to the loss of stability of the coordination state near antiphase, π rad. The flow now belongs to the monostable régime. Following

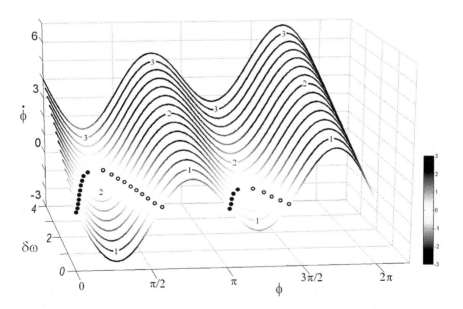

Fig. 3 *Elementary coordination law* (Eq. 1). *Surface formed by a family of flows of the Order Parameter or coordination variable ϕ (in radians) as a function of $\dot{\phi}$ for increasing values of $\delta\omega$ between 0 and 4. For this example, the coupling is fixed: $a = 1$ and $b = 1$ (see text for details)*

line 3 from left to right, no stable or unstable fixed points exist yet coordination has not completely disappeared. This flow corresponds to the *metastable* régime, a subtle blend of coupling and intrinsic differences between the parts in which behavior is neither completely ordered (synchronized) nor completely disordered (desynchronized). *It is the subtle interplay between the coupling (b/a) and the symmetry breaking term* $\delta\omega$ *in Eq. 1 that gives rise to metastability.*

Equation 1 is somewhat odd. Even though it is an order parameter equation of motion that is designed to describe the evolution of *collective behavior* (in words, phi dot is a function of phi), it includes also a parameter $\delta\omega$ that arises as a result of differences (*heterogeneity*) among the *individual components*. Equation 1 is thus a strange mixture of the whole and the parts, the global and the local, the cooperative and the competitive, the collective and the individual. Were the components identical, i.e., no diversity, $\delta\omega$ would be zero and we would not see component differences affecting the behavior of the whole. Equation 1 would simply reflect the behavior of the collective untarnished by component properties, a purely emergent interaction—the HKB equation. It is the fact that *both* the components *and* their (nonlinear) interaction appear at the same level of description that gives rise to the array of coexisting tendencies characteristic of metastability. Equation 1 is a basic representation of a *synergy* or *coordinative structure*: a low dimensional dynamic of a multi- and metastable organization in which the tendency of the parts to act together coexists with a tendency of the parts to do their own thing (Kelso 1995, Ch 4; for more on synergies, see Kelso 2009a, b). It is metastability that endows the synergy with robustness and flexibility, enabling the same parts to participate in multiple functions. If the synergy is a unit of life, then it is metastable dynamics that brings it alive. We'll come back to this point and its broader implications in the final section of the paper.

On Learning and the Nature of Change. So far, Coordination Dynamics (CD) provides an empirically validated theoretical account of what "working together" means and places the functional synergy on the pedestal of biological coordination. What does the CD say about *learning*? Only a few brief remarks can be made here (but see Kostrubiec et al. 2012 for a review of empirical and theoretical modeling work on learning conducted with Pier-Giorgio Zanone and colleagues in Toulouse over a period of 25 years). Coordination Dynamics defines *learning as the modification of a pre-existing repertoire that is unique to each individual.* Thus, in CD the *individual is the significant unit of analysis*; every individual enters the learning situation with their own biases/predispositions/coordination tendencies. This individual signature (which must be quantified) is referred to as *intrinsic dynamics*. Here again, the dynamics refer to the dynamics of collective variables that span both the organism and the environment. What changes during learning? Experiments show that not just the pattern to be learned changes during learning, *but the entire landscape of the intrinsic dynamics/pre-existing repertoire.* What is the nature of change? In his *Autobiography*, Charles Darwin, based on observations of his own children, concluded that changes due to learning "have all had a gradual and natural origin". Our results show that *learning can be smooth and continuous or abrupt and qualitative depending on the relationship* *between the learner's pre-existing repertoire and the new information that is to be learned. Competition* between the pre-existing repertoire and new information is a key mechanism that dictates the nature of change. Stability, not just error correction is the overall criterion for learning. Indeed, the brain areas recruited for learning and their level of activation are directly related to the stability of performance (Jantzen et al. 2009; DeLuca et al. 2010). Bottom line?

According to Coordination Dynamics, if you want to change anything and have it persist "permanently" (as opposed to being a mere transient)—in other words *learn*—you'd better know the system's intrinsic dynamics. Knowing the latter means you know *what* to modify, and whether to use competitive or cooperative mechanisms to cause abrupt or gradual change (see Fig. 4). I suspect this principle of learning operates at all levels, from individuals through society and is at the heart of significant political change. Foreign policy, diplomacy and acts of aggression often flounder because of ignorance about the intrinsic dynamics of the system that they aim to influence or change. Obtaining measures of intrinsic dynamics in all these situations constitutes a major challenge—though with major payoffs because it means you know what to change. Economically speaking, as the pharmaceutical business has begun to appreciate, knowing the intrinsic dynamics of the individual is at the core of so-called personalized medicine—for example in understanding why a drug has a positive effect on one person and no effect on another. Statistical studies in clinical populations hide this fact.

On Agency. *Learning to live together requires purpose and intent.* There must be a desire to make things happen. Though grounded in evolutionary and self-organizing processes, synergies or coordinative structures are meaningful and goal-directed. Working and living together is not simply the result of a reflexive herd or group instinct: it requires agency. Agency means action toward an end. So where do agency and goal-directedness come from? A main aspect of self-organizing dynamical systems is that the emergence of pattern and pattern switching occur *spontaneously,* solely as a result of the dynamics of the system: no specific ordering influence from the outside and no homunculus-like agent or program inside is responsible for the behavior observed. Yet somehow, that is, without magic or some vital force, what we call agency must spring from the ground of spontaneous self-organized activity (Kelso 2002). A clue comes from studies of 3 month-old human babies (Rovee and Rovee 1969). When babies are comfortable and lying in their crib they kick their legs and move their arms spontaneously. After a while they become fussy and so to amuse them mothers will sometimes attach a mobile above their head that looks attractive and makes noises that babies seem to like. But this doesn't last forever. Maybe the baby gets bored or attention toward the mobile saturates. What if you tie a ribbon to the baby's ankles and attach it to the mobile hanging over the baby's crib? By virtue of the coupling, any spontaneous foot or leg movements will cause the mobile to move. Now look what happens next. Suddenly the baby increases his/her kicking rate by a factor of 4! The baby realizes that it, not some outside force, is making the mobile move! The idea is that when the baby perceives it is causing the world to change it experiences itself as an agent for the very first time.

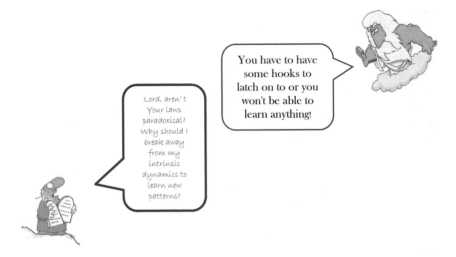

Fig. 4 *One of the messages from research on the coordination dynamics of learning* (thanks to Vivianne Kostrubiec for cartoon)

It seems that the key to the emergence of conscious agency is not only spontaneous movement (which is a fundamental component, nevertheless) but the bidirectional coupling (by means of the tether) between the baby and the world. Theoretically speaking, a coordinative structure *qua* coupled dynamical system is formed when the (notably prelinguistic) infant discovers itself as an agent ('this is me'), that is, when the baby realizes it can *make things happen*. In this theory (Kelso 2016; Kelso and Fuchs 2016) the birth of agency and its causative powers ("I do", "I can do") corresponds to a *phase transition* of a coordination dynamics whose key variables span the interaction between the organism (baby) and its environment (the moving mobile). This igniting of agency has a eureka-like, 'aha' effect; mathematically, it corresponds to a bifurcation in the coupled dynamics. Here, coupled dynamics refers to the coordinated *relation* between the baby's movements and the (kinesthetic, visual, auditory and emotional) consequences they produce. Bifurcations are the mathematical equivalent of phase transitions, qualitative changes in coordinative states. The main mechanism underlying the origin of self as a causal agent involves *positive feedback*: when the baby's initially spontaneous movements cause the world to change, their perceived consequences have a sudden and sustained amplifying effect on the baby's further actions. This autocatalytic mechanism is continuous with our understanding of how biological form develops and of the feedforward network motifs so ubiquitous in the design of biological circuits (Alon 2007). The deep irony of this theory of the coordination dynamics of moving bodies is that the most primitive form of self-organization known in biological coordination (brains included), a synergetic phase transition, gives rise to self. The root soil of agency, as Sheets-Johnstone (Sheets-Johnstone 1999/2011) would say, rests on primal animation, on being alive and moving.

Metastable Mind—The key to learning to learning to live together? What are the implications of multi- and metastable coordination dynamics (cf. Fig. 3) for understanding the mind? Like nature and nurture, the contents of the mind and the dynamics of the mind are inextricably connected. Thoughts are not static: Like the flow of a river, they emerge and disappear as patterns in a constantly shifting dynamic system (Kelso 1995). Though this is a nice metaphor, science demands we go beyond it to seek description and explanation. In particular, we would like to explain or understand the brain $\sim$ mind relation—if possible–with a single theoretical model. Figure 5 is intended to convey the gist of the story. On the left side of the middle panel, two areas of the brain (for the sake of simplicity) are shown to be active. This acknowledges a simple fact—or at least a dominant assumption in contemporary neuroscience: The contents of thoughts depend on the neural structures activated. However, identifying thought-specific structures and circuitry using brain mapping, important though it may be, is hardly sufficient to tell us how *thinking* works. Active, dynamic processes like perceiving, attending, remembering and deciding that are associated with the word "thinking" are not restricted to particular brain locations, but rather emerge as patterns of interaction in time among widely distributed neural ensembles, and in general between human beings and their worlds.

One of the great riddles of contemporary neuroscience is how the multiple, diverse and specialized areas of the brain are coordinated to give rise to thinking and coherent goal-directed behavior. A key fact embraced by Coordination Dynamics is that neuronal assemblies in different parts of the brain oscillate at different frequencies. Such oscillatory activity is a prime example of self-organization in the brain. But oscillation, though necessary is not sufficient. It is, rather, that oscillations are coupled or "bound" together into a coherent network when people attend to a stimulus, perceive, remember, decide and act (e.g., Başar 2004; Bressler and Kelso 2001; Buzsáki 2006; Kelso 1995; Varela, et al. 2001; Singer 2005, for reviews). This is a dynamic, self-assembling process, parts of the brain engaging and disengaging in time, as in a proverbial country square dance. In the simplest case shown in the left column of Fig. 5, oscillations in different brain regions can lock "in-phase", brain activities rising and falling together, or "anti-phase", one oscillatory brain activity reaching its peak as another hits its trough and vice versa. In-phase and antiphase are just two out of many possible multistable, phase attractive states that can exist between multiple, different, specialized brain areas depending on their respective intrinsic properties and functional connectivity. More broadly, the organism and its environment are embedded in a nested frame of rhythms ranging from rest $\sim$ activity and sleep cycles to circadian and seasonal rhythms that both modify and are modified by behavior, development and aging.

The top left part of Fig. 5 conveys the essential *bistable* nature of brain $\sim$ mind. Two states are possible for identical parameter values: which state one enters depends on initial and boundary conditions. According to Coordination Dynamics, *bistability is the basis of polarization and the either/or*. Note that this does not necessarily imply any judgment of good or bad. Polarization, for example, may be

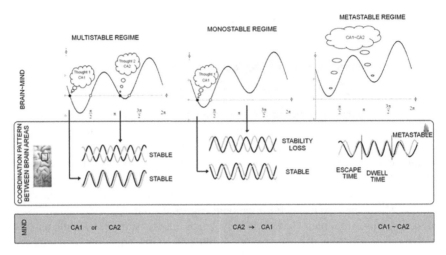

Fig. 5 *Elementary Coordination Dynamics of Brain ~ Mind.* Middle panel represents synaptically coupled brain oscillations from two brain areas (for the sake of simplicity) whose activation is meaningful and specific to the content of "thoughts". Here "thought" is used in a generic sense; the states could refer to patterns of perceiving, emoting, remembering, deciding, acting, etc. Top left panel shows the layout of the fixed points of the relative phase dynamics (Eq. 1) in the multi- (here bi-)stable regime. Solid circles are stable and attracting; open circles are unstable and repelling (see also Fig. 3). Two states are stable corresponding to particular phase relations between groups of neurons/brain areas, representing two stable "thought" patterns (ca1 and ca2) for exactly the same parameter values. Top middle panel shows that the formerly stable pattern near antiphase switches to near inphase as a result of changing circumstances. Any ambiguity due to bistability has been removed, a "decision" or "selection" has been made and as a result, the system is monostable and confined to a single thought pattern. The switching mechanism is dynamic instability induced by changing control parameters (e.g., the coupling between the neural populations which may be altered by neuromodulators). Fluctuations (not explicitly represented here) also play a key role in spontaneous switching. Top right panel shows that all states, both stable and unstable have disappeared. This is the metastable régime. Now "thoughts" no longer correspond to fixed point, fully synchronized states of the coordination dynamics, but rather to coexisting tendencies or dispositions that have characteristic dwell times. The lowest panel called "Mind" illustrates the classical dual nature of either/or, binary oppositions between complementary aspects (ca1 *or* ca2), mind (and mindset) switching (ca2 to ca1 and *vice versa*) and the mind and mindset of the complementary nature, where both "thoughts" are held in the mind at the same time (ca1 ~ ca2) (adapted from Kelso 2008)

seen as the driving tension behind scientific progress in the sense of Thomas Kuhn (1962), and bistability may be exploited for solving ill-defined problems where the consideration of multiple interpretations of data is an advantage. Bistable, and in general multistable coordination dynamics confers many advantages on living things, in particular multifunctionality (see, e.g., Kelso 1991).

Coordination dynamics suggests that the persistence of a thought depends on the *stability* of the brain's relative phase dynamics. Some thoughts persist longer than others because the phase relations underlying them are more stable. In Fig. 5 (top left), the negative slope through the ordinate near in-phase ("thought 1") is greater, hence more stable, than its anti-phase counterpart ("thought 2"). This proposition is supported by experiments and specific neurally-based modeling which shows that different patterns of spatiotemporal brain activity are differentially stable.

So what causes thoughts to switch, the person to change her mind? The middle column of Fig. 5 offers a specific mechanism: dynamic instability. Considerable experimental evidence has demonstrated that switching in both brain and behavior is a self-organized process that takes the form of a nonequilibrium phase transition. Fluctuations play a key role, testing the stability of states and enabling the system to discover new states. In Coordination Dynamics, once the system settles into an attractor, a certain amount of noise or a perturbation is required to switch it to another attractor. Or, if internal or external conditions change when the system is near instability, a bifurcation or phase transition may occur, causing the system to switch from being multistable to monostable or vice versa (see Ditzinger and Haken 1989, 1990 for excellent examples of such modeling). Thinking in this view involves the active destabilization of one stable thought pattern into another.

A slightly different view emerges from the flow of the relative phase dynamics in the metastable régime (Fig. 5, right). Instead of thoughts corresponding to phase synchronized states in the brain that must be destabilized if switching is to occur, metastability consists of a more subtle dwell and escape dynamic in which a thought is never quite stable and merely expresses a joint *tendency* for neural areas to synchronize together and to oscillate independently. Fluid thinking, in this view, is when the brain's oscillations are neither completely synchronized nor desynchronized (see also area 2 in Fig. 3). In the metastable régime, successive visits to the remnants of the fixed points are intrinsic to the time course of the system, and do not require any external source of input. Switching occurs, of course, but continuously and without the need for additive noise or changes in parameters. From the perspective of Coordination Dynamics, the time the system dwells in each remnant depends on a subtle blend of the asymmetry of the rhythmic elements (longer dwelling for smaller asymmetry) and the strength of the coupling (longer dwelling for larger values of a and b in Eq. 1). Metastable coordination dynamics also rationalizes William James (1890) beautiful metaphor of the stream of consciousness as the flight of a bird whose life journey consists of 'perchings' (phase gathering, integrative tendencies) and 'flights' (phase scattering, segregative tendencies). Both tendencies appear to be crucial: the former to summon and create thoughts; the latter to release individual brain areas to participate in other acts of cognition, emotion and action.

Metastability (meta meaning beyond) is an entirely new conception of brain organization, not merely a blend of the old. Individualist tendencies for the diverse regions of the brain to express their independence coexist with coordinative tendencies to couple and cooperate as a whole. As we have seen, in the metastable brain local segregative and global integrative processes coexist as a complementary pair, not as conflicting theories. Reducing the strong hierarchical coupling between the parts of a complex system while allowing them to retain their individuality leads to a looser, more secure, more flexible form of functioning that promotes the creation of information. Too much autonomy of the component parts means no chance of them coordinating and communicating together. On the other hand, too much interdependence and the system gets stuck, global flexibility is lost. Well-known manifestations of too much synchronization in the brain, for example, are characteristic of diseases like Parkinson's disease and epilepsy.

Metastability is an expression of the full complexity of brains and people (Kelso 2001; Tognoli and Kelso 2014) and gives rise to a plethora of complementary pairs. In fact, it is chock full of them (Kelso and Engstrom 2006):

individual $\sim$ collective
parts $\sim$ whole
segregation $\sim$ integration
choice $\sim$ chance
competition $\sim$ cooperation
symmetry $\sim$ broken symmetry
stability $\sim$ instability
states $\sim$ tendencies/dispositions
and so forth.

The tilde ($\sim$) or squiggle symbol expresses a basic truth: both members of a complementary pair are required for understanding. One without the other is incomplete. Even polarization $\sim$ reconciliation is a complementary pair.

It is a truism that we live in a world of 'isms'. Such distinctions have no doubt served a (sometimes malicious) purpose (think, e.g. of the term 'birtherism'). The problem is that we human beings—intentionally or not—have reified them. We hear a lot these days, for example, of the need to replace reductionism with emergentism—sometimes accompanied by fond hopes that this step will pave the way to a new global civilization. Often there is little or no scientific basis for promulgating a new worldview. "Isms" are an obstacle to understanding: they tend to result in one doctrine being defended over another rather than opening up new ideas.

Dynamically speaking, bistability is at the core of polarization and the either/or, the latter often posed as "isms". But as can be seen in Figs. 3 and 5, that way of thinking is just a (very) restricted régime of the coordination dynamics.

Metastability is telling us that it's not just about states and it's not just about one tendency or the other, but both tendencies at the same time. *Contraria sunt complementa* as Niels Bohr's famous coat of arms says, an attitude hearkened to also by "Einstein's conscience", the great Wolfgang Pauli:

> *To us the only acceptable point of view is one that recognizes both sides of reality—the quantitative and the qualitative, the physical and the psychical—as compatible with each other, and can embrace them simultaneously.* (Pauli 1952)

Inna Semetsky (2010), an eminent leader in education based at the Institute of Advanced Study for Humanity at the University of Newcastle in Australia, draws attention to disciplinary knowledge based on the classical logic of the excluded middle—where subject and object, self and other are separated. In seeking a transdisciplinary, in vivo knowledge system that connects subject and object, self and other she and others advocate a logic of the *included* middle, represented by the squiggle symbol of The Complementary Nature. Semetsky makes a strong case for non-dualistic transdisciplinary knowledge as being at the core of a new *ecoliteracy* movement based on sharing and cooperating with others and the inclusion of values. Facts alone will not do. For her, the task of transforming human structures into open ended ecological systems in harmony with the natural world represents a major challenge at all levels, most especially at the level of education. Here again, the complementary organism ∼ environment pair, central to Coordination Dynamics—"the intercourse of the live creature with its surroundings", as Semetsky quotes Dewey as saying—is central also to ecoliteracy.

It seems that the message of metastable coordination dynamics is that there is a deep principle of complementarity underlying life, brain, mind and society. Metastability says that complementary aspects and their dynamics are found not just at the remote level of subatomic processes dealt with by quantum mechanics, or even at the level of molecular complementarity as in DNA, but at the level of human beings, human brains and human behavior–at the level, in other words, where the science of coordination plays out—Coordination Dynamics. Thinking narrowly in terms of contraries and the either/or is easy when life is simple. But in complex coordinated systems it seems that sharp dichotomies and contrarieties have to be replaced with far more subtle and sophisticated complementarities. One suspects this is true for all of nature, human nature (and human brains) included. Understanding ourselves and the way we think and behave in our relationship to the environment, to other species or relationships with others, individually as human beings, as groups, as societies, as nation states, would seem to depend on this realization.

Acknowledgements The author's research is currently supported by National Institute of Mental Health grant MH080838, the Davimos Family Endowment for Excellence in Science, and the Florida Atlantic University Foundation (Eminent Scholar in Science).

References

Alon, U. (2007). *An introduction to systems biology: Design principles of biological circuits.* Chapmann & Hall/CRC.

Başar, E. (2004). *Memory and brain dynamics: Oscillations integrating attention, perception, learning, and memory.* Boca Raton: CRC Press.

Bressler, S. L., & Kelso, J. A. S. (2001). Cortical coordination dynamics and cognition. *Trends in Cognitive Sciences, 5,* 26–36.

Buzsáki, G. (2006). *Rhythms of the brain.* Oxford: Oxford University Press.

Corning, P. A. (2010). Rotating the Necker cube: A bioeconomic approach to cooperation and the causal role of synergy in evolution. *Journal of Bioeconomics, 15,* 171–193.

DeLuca, C., Jantzen, K. J., Comani, S., Bertollo, M., & Kelso, J. A. S. (2010). Striatal activity during intentional switching depends on pattern stability. *Journal of Neuroscience, 30*(9), 3167–3174.

Ditzinger, T., & Haken, H. (1989). Oscillations in the perception of ambiguous patterns. *Biological Cybernetics, 61,* 279–287.

Ditzinger, T., & Haken, H. (1990). The impact of fluctuations on the recognition of ambiguous patterns. *Biological Cybernetics, 63,* 453–456.

Fuchs, A., & Kelso, J. A. S. (2017). Coordination dynamics and synergetics: From finger movements to brain patterns and ballet dancing. In S. Mueller et al. (Eds.), *Complexity and Synergetics* (pp. 301–316). Heidelberg: Springer-Verlag.

Haken, H. (1977). *Synergetics, an introduction: Non-equilibrium phase transitions and self-organization in physics, chemistry and biology.* Berlin: Springer.

James, W. (1890). *The principles of psychology* (Vol. 1). New York: Dover.

Jantzen, K. J., Steinberg, F. L., & Kelso, J. A. S. (2009). Coordination dynamics of large-scale neural circuitry underlying sensorimotor behavior. *Journal of Cognitive Neuroscience, 21,* 2420–2433. https://doi.org/10.1162/jocn.2008.21182.

Kelso, J. A. S. (1991). Behavioral and neural pattern generation: The concept of Neurobehavioral Dynamical System (NBDS). In H. P. Koepchen & T. Huopaniemi (Eds.), *Cardiorespiratory and motor coordination.* Berlin: Springer-Verlag.

Kelso, J. A. S. (1992). Coordination dynamics of human brain and behavior. *Springer Proc. in Physics, 69,* 223–234.

Kelso, J. A. S. (1995). *Dynamic patterns: The self-organization of brain and behavior.* Cambridge, MA: The MIT Press (Paperback edition, 1997, 4th Printing).

Kelso, J. A. S. (2001). Metastable coordination dynamics of brain and behavior. *Brain and Neural Networks (Japan), 8,* 125–130.

Kelso, J. A. S. (2002). The complementary nature of coordination dynamics: Self-Organization and the origins of agency. *Journal of Nonlinear Phenomena in Complex Systems, 5,* 364–371.

Kelso, J. A. S. (2008). An essay on understanding the mind. *Ecological Psychology, 20,* 180–208.

Kelso, J. A. S. (2009a). Coordination dynamics. In R. A. Meyers (Ed.) *Encyclopedia of complexity and system science,* Springer: Heidelberg (pp. 1537–1564).

Kelso, J. A. S. (2009b). Synergies: Atoms of brain and behavior. *Advances in Experimental Medicine and Biology, 629,* 83–91. ((Also D. Sternad (Ed) *A multidisciplinary approach to motor control.)* Springer, Heidelberg).

Kelso, J. A. S. (2012). Multistability and metastability: Understanding dynamic coordination in the brain. *Philosophical Transactions Royal Society B, 367,* 906–918.

Kelso, J. A. S. (2016). On the self-organizing origins of agency. *Trends in Cognitive Sciences, 20* (7), 490–499. https://doi.org/10.1016/j.tics.2016.04.004.

Kelso, J. A. S., & Engstrom, D. A. (2006). *The complementary nature,* Cambridge, MA: The MIT Press. Paperback Edition.

Kelso, J. A. S., & Fuchs, A. (2016). The coordination dynamics of mobile conjugate reinforcement. *Biological Cybernetics, 110*(1), 41–53. https://doi.org/10.1007/s00422-015-0676-0.

Kelso, J. A. S., Dumas, G., & Tognoli, E. (2013) Outline of a general theory of behavior and brain coordination. *Neural Networks,* 37, 120–131. (25th Commemorative Issue).

Kostrubiec, V., Zanone, P.-G., Fuchs, A., & Kelso, J. A. S. (2012). Beyond the blank slate: Routes to learning new coordination patterns depend on the intrinsic dynamics of the learner—experimental evidence and theoretical model. *Frontiers in Human Neuroscience, 6,* 212. https://doi.org/10.3389/fnhum.2012.00222.

Kuhn, T. S. (1962). *The structure of scientific revolutions.* Chicago: University of Chicago Press.

Latash, M. (2008). *Synergy.* Oxford, New York: Oxford University Press.

Maynard-Smith, J., & Szathmáry, E. (1995) *The major transitions in evolution.* Freeman.

Noble, D. (2008) *The music of life.* Oxford University Press.

Oullier, O., DeGuzman, G. C., Jantzen, K. J., Lagarde, J., & Kelso, J. A. S. (2008). Social coordination dynamics: Measuring human bonding. *Social Neuroscience.* https://doi.org/10.1080/17470910701563392. First Published on: 12 October 2007.

Pattee, H. H. (1976). Physical theories of biological coordination. In M. Grene & E. Mendelsohn (Eds.) *Topics in the philosophy of biology,* Vol.27, pp. 153–173.

Pauli, W. (1952/1994). *Writings on physics and philosophy*. Heidelberg: Springer.

Rovee, C. K., & Rovee, D. T. (1969). Conjugate reinforcement of infant exploratory behavior. *Journal of Experimental Child Psychology 8*, 33–39.

Semetsky, I. (2010). Ecoliteracy and Dewey's educational philosophy: Implications for future leaders. *Foresight, 12*, 31–44.

Sheets-Johnstone, M. S. (1999/2011). *The primacy of movement*, John Benjamins.

Singer, W. (2005). The brain—An orchestra without a conductor. *Max Planck Research, 3*, 15–18.

Stergiou, N. (this volume).

Tognoli, E., & Kelso, J. A. S. (2014). The metastable brain. *Neuron, 81*, 35–48.

Varela, F. J., Lachaux, J.-P., Rodriguez, E., & Martinerie, J. (2001). The brainweb: Phase synchronization and large-scale integration. *Nature Reviews Neuroscience, 2*, 229–239.

Zhang, M., Kelso, J. A. S., & Tognoli, E. (2018). *Critical diversity: United or divided states of social coordination*. https://doi.org/10.1371/journal.pone.0193843.

Part III
The Brain's Trust

The Neurobiology of Reward: Understanding Circuitry in the Brain that Shapes Our Behavior

Chris Evans

Species evolution has been fueled by the development of brain neurocircuitry and physical attributes to compete for food, partners and environmental niches. This neurocircuitry that was designed to respond to natural rewards has been the subject of considerable research effort given its importance to motivational behavior. Reward-related behaviors are determined by a complex array of variables including how good we feel, how much we like or want a reward, competition with other rewards, reaction to the absence of a reward, societal values and accessibility to the reward—responses regulated by current and past experiences interacting with our genetic makeup. Understanding the neurobiology of hedonic tone (how good we feel) and the circuitry behind reward-motivated actions (how they are acquired, retained and regulated), particularly in contrast to habitual actions is essential, particularly as these are activities that substance abuse has provided an unfortunate model. This presentation will explore the reward circuitry of the brain and focus on how drug-taking and other repeated reward behaviors may reduce hedonic tone and switch decision-making brain circuitry away from executive areas of the cortex reflecting the metaphorical switch from casual drug-user to addict (Leshner 1997).

C. Evans (✉)
Brain Research Institute, University of California, Los Angeles, CA, USA
e-mail: cevans@ucla.edu

© Springer International Publishing AG, part of Springer Nature 2019
J. A. S. Kelso (ed.), *Learning To Live Together: Promoting Social Harmony*,
https://doi.org/10.1007/978-3-319-90659-1_11

The Importance of Brain Reward Circuitry

In single cell organisms such as amoeba and paramecium, cellular processes that are necessary for survival have evolved to detect, engulf and digest food, propagate via cell division, and avoid harm. As single cells evolved into animals, the coordination of cells specialized for detection, decision-making and executing actions was required to optimize survival. As early in evolution as c-elegans, worms diverging from the vertebrate lineage in the pre-Cambrian era about 1 billion years ago, dopamine signaling via 8 neurons has been found essential for food foraging behaviors and acquiring appropriate behavioral responses to tap avoidance (Schafer 2004). In the case of vertebrates, including humans and other primates, specialized regions of the brain, some, like worms, signaling via dopamine, are responsible for motivating and focusing behavioral functions associated with reward and aversion. Evolution has built an extremely powerful neuronal engine in the human brain to ensure individual and species survival and managing this horsepower with current societal values that require extensive modulation of these primitive survival-associated pathways is an extreme challenge. At both a societal and individual level there is clear evidence that circuitry modulating reward behaviors can readily become dysfunctional and so understanding these systems and how they are regulated and educated is essential.

Identification of Brain Areas Modulating Reward and Aversive Behavior

Experiments by Dr. Jim Olds in the early 1950's identified structures in the brain of rodents he called "pleasure centers". When electrical brain stimulation was self-delivered by pressing a lever, electrical impulses delivered into several sub-cortical areas associated with dopaminergic pathways resulted in up to 2000 lever presses per hour—truly compulsive lever pressing! Indeed, the rats would stimulate these brain regions to the exclusion of eating when hungry or even caring for offspring and would endure extremely unpleasant situations to receive the brain stimulation. Similar "pleasure centers" have been identified in all mammals including humans and have been mapped in great detail in rodents (Milner 1991). Interestingly, the mapping reveals that many different and, in some cases, nominally interconnected brain areas can elicit self-stimulation, whilst other areas of the brain when stimulated cause aversion. What's striking about these experiments is the discovery that stimulation of very discrete areas of the brain can trump all other input and completely control behavioral responses, demanding the animal repeat or avoid stimulation to the exclusion of any other behavior. The electrical stimulation clearly activates neuronal circuitry in the brain. The question of which neurotransmitters and which neurons has been the focus of considerable research and these questions have answers, at least in mesolimbic areas of the brain that are related to reward.

Neurotransmitters in the Brain Associated with Reward

Electrical self-stimulation of the mesolimbic dopamine pathway connecting the ventral tegmental area to nucleus accumbens is exceptionally robust. Since disruption of dopamine neurotransmission disrupts self-stimulation, an association of dopamine with "pleasure" and the reward pathways has long been established (Milner 1991). Increase of synaptic dopamine via drugs such as cocaine or amphetamine, self-administered via lever pressing directly into the nucleus accumbens parallels electrical stimulation in that obsessive lever pressing can be observed. All of the addictive drugs that have been assessed trigger dopamine release in the nucleus accumbens, some drugs such as cocaine and amphetamine working directly on the dopamine neurons, others such as opiates, alcohol and cannabinoids work by disinhibition of the GABAergic system. It is an accepted dogma that drugs of abuse require dopamine release in the nucleus accumbens to exhibit their "rewarding" properties although alternative pathways for self—administration of certain drugs have been established. Genetic studies in mice have revealed an interdependent synergy between different endogenous systems having a reward-like action (Gaveriaux-Ruff and Kieffer 2002). For example, the endogenous opioid system appears a key player and if disrupted by genetically removing opioid receptors from the mouse genome, the rewarding effects of alcohol, nicotine and cannabis, which have primary sites of action at receptors other than opioid receptors, are negated. Naltrexone, an antagonist of opioid receptors that blocks the effects of endogenous opioids has been shown to have clinical efficacy in the treatment of alcohol addiction. Furthermore, if the cannabinoid system is disrupted opiate drug reward is compromised and cannabinoid antagonists are now in clinical trials for both nicotine addiction and obesity. The bottom line is that drugs of abuse have provided extremely valuable tools for identifying systems, circuits, cells and molecules that are involved in reward circuitry.

Brain Circuitry Regulating Reward-Mediated Behaviors

There are many processes requiring consideration when assessing reward—mediated behavior. One initial process is "liking" or "disliking" a judgment that, as we are all well aware, can be highly variable among individuals. For example, some folk are in heaven when given opiate drugs, others may detest them and yet both reactions are mediated by the same physiological effect. Other reasons for individual responses may be differential intensity of aversive side effects such as racing heart in the case of cocaine or nausea in the case of opioids. Even when there are no obvious differences in perception, responses to liking or disliking specific drugs, food-related, reproductive and nurturing behaviors and even pain-elicited responses differ immensely from person to person. What causes these differences is likely a complex interaction of genetics with environmental experience and societal priming. Work using sugar as a reward for rodents has been a primary tool to study brain areas involved in liking. Based on lesion studies and the observation that opioids enhance taste hedonic reactions in humans and rodents, Kent Berridge and colleagues have identified two brain areas involved in opioid-enhanced sugar liking, namely the caudal ventral pallidum and the shell of the nucleus accumbens (Pecina et al. 2006). Experiments use microinjections of opioids into discrete areas of the brain to enhance facial expressions correlated with pleasure from the sugar. Whether the same responses would be observed in subjects that dislike opioid drugs is questionable and, indeed, it is probable that liking of other sensation modalities will be generated by different areas of the brain. Nevertheless, a study of a patient with bilateral ischemic damage to the ventral pallidum offers broad support for the involvement of this area in hedonic mechanisms. Prior to the damage the patient had a long history of substantial polysubstance abuse. Following the damage, presumed to have been caused by psychostimulants, he was depressed, there was no drug craving and he no longer found drinking alcohol pleasurable (Miller et al. 2006). Hedonic tone in mice is markedly in influenced by the endogenous opioid system. Mice given the opioid receptor antagonist naltrexone in a specific environment reliably avoid that environment when subsequently given access to alternatives. Using mice with genetic ablation of components of the endogenous opioid system, Nigel Maidment and colleagues at UCLA have determined both the opioid peptides and the opioid receptor involved (Skoubis et al. 2005) and studies using microinjection of opiate antagonist into discrete brain areas have implicated the ventral pallidum as the likely site of action (Skoubis and Maidment 2003).

Dopamine neuronal projections in the mesolimbic pathway appear critical not only for self-stimulation but also the effects of many drugs of abuse as well as natural "rewards" such as food and drink, so they have long been considered a primary gateway for reward (Wise 2004). However, the exact role of dopamine has been somewhat controversial with regard to the involvement in the liking or hedonic value as opposed to wanting or driving the anticipation of reward, with evidence suggesting it is more closely involved in the latter than the former. Regardless of its precise role, the mesolimbic dopamine pathway is a key player in triggering corticostriatal pathways that are the decision makers for reward behaviors. Two types of behaviors arc defined; goal-directed behaviors i.e. behaviors driven by obtaining a reward for its hedonic value, and habitual behaviors which occur automatically with minimal attention to the reward itself or, in simple terms, are determined by a stimulus-response connection without much analysis of the goodies. After continued learning and rehearsal of goal-directed actions for a reward, these actions can become habitual; the behavior becomes automatic, unrelated to the goals and very difficult to reverse. Habitual behaviors are associated with many rewarding stimuli in our society particularly evident in the case of eating and drug taking. The transition of goal-directed behaviors to habitual behaviors can be modeled by sophisticated instrumental behavioral paradigms in rodents. The studies by Balleine and colleagues at UCLA have over the years used such paradigms to dissect brain regions and pathways involved in the transition from goal—directed to habitual behavior (Balleine 2005; Balleine et al. 2007). What is clear from these experiments, which use food as a reward, is that the transition from the goal-directed to the habitual responses involves circuitry switching. Goal-directed behavior is mediated by corticostriatal circuits involving the dorsomedial striatum and its afferents in the medial prefrontal cortex (the executive processing area of the cortex), whereas habitual behaviors shift decision-making to the dorsolateral striatum and sensory motor cortex (an area lacking in analytical processing). What is important about this circuitry switch is that habitual actions occur with less thought—the control of behavior transitions to areas of the brain automatically responding to cues without extensive executive processing. The prefrontal cortex is a key cortical structure involved in inhibiting or constraining the actions of the more primitive limbic-striatal circuitry—it is considered the area for executive function and some argue it is the extensive elaboration of this brain structure in humans that differentiates us from other primates. In habit-acquired behaviors, the pre-frontal cortex is removed from the decision-making circuit thus reducing the possibility to modify or control the behavior—the switch is set to go on cue without thinking. Several functional neuroimaging studies in the brains of addicts observe deficient usage of the pre-frontal cortex and associated areas adding support to a shifted circuitry in processing following repeated drug taking (Aron and Paulus 2007).

Lessons from Drug Abuse

By now everyone is acutely aware that smoking is a major cause for lung cancer and circulatory disease. Smoking cigarettes accounts for approximately 15% of preventable deaths in the US (approx. 400,000 persons/year), overeating accounts for just under 20% of deaths and drinking alcohol is responsible for about 5% of deaths, mainly from liver sclerosis and accidents. Despite knowledge of these unhealthful outcomes 20–40% (depending on the age group) of the US adult population smoke cigarettes, 5–15% drink heavily and fatty unhealthful foods continue to be over consumed. Other illicit drug use, particularly marijuana and addictive prescription medicines is also widespread with approximately 5–15% (depending on the age group) of the population regularly using. In looking at other measures of deviant reward-seeking behavior, sexual abuse remains prevalent and even though vastly underreported, there was close to 200,000 victims of rape or sexual assault in the US in 2005. On another side of the hedonic coin, depressed mood is prevalent with about 10% of the adult population suffering each year. In 2004 there were over 800,000 attempts of suicide reported with suicide accounting for about 1.5% of deaths in the US. The key deduction from these data is that a substantial proportion of the population are, in one way or another, battling issues concerning the regulation of pathways in the brain encoding hedonic tone and reward. Drugs of abuse are troubling because they rewire the reward system rendering behavioral control of maladaptive actions more difficult. Common symptoms in an addict are desensitized appreciation of natural rewards combined with a diminished reward value of the abused drug. It is likely that decision-making processes surrounding behavior to obtain the drug, as with food reward, gradually switches from goal-directed pathways to predominantly habit pathways that exclude the executive processing via the prefrontal cortex. Furthermore, many drugs of abuse appear to engage opponent processes that, as the name suggests, oppose the primary effects of the drug. This can be disastrous and drive addiction via creating the opposite behavioral effects to those for which the drug was taken. Opponent processes are very clear in the case of opioids where they are used for euphoria, analgesia, or constipation and when stopped (withdrawal) cause the opposite effects namely dysphoria, hyperalgesia and diarrhea (Bryant et al. 2005).

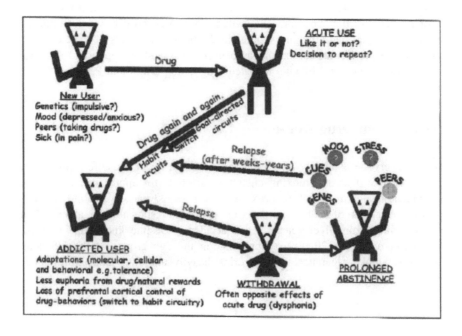

One prominent opinion is that hedonic tone is continually decreased by drug taking as a consequence of opponent processes and that this is a major driver of drug taking and addiction; the drug temporarily rectifies the dysphoria (Koob 1992). The figure above indicates pathways to stages in drug addiction and depicts mechanisms that may underlie addictive behaviors. The decision to take a drug is complex and deserves much research effort in order to tackle abuse at its roots. It is true that not all people who experiment with drugs become abusers and it is likely that genetic and environmental factors work together to create susceptibility—perhaps related to the ease of switching to habit circuitry. Genetics undoubtedly plays a large role in susceptibility to substance abuse (Hiroi and Agatsuma 2005) and some estimates from twin studies have determined that as much as 60% of susceptibility is due to inheritance. Once addicted it is often extremely difficult to quit. Relapse rates for users of heroin are close to 100% hence maintenance therapy using methadone or buprenorphine is often recommended in order to stabilize the addicted brain. Exciting new research is emerging that is suggestive of a possible re-education strategy called reconsolidation whereby memories that are triggered and active become somewhat labile and plastic for remolding (Alberini et al. 2006). However, at present the mechanisms for remolding are somewhat drastic and further research is clearly warranted. Government policies on prevention have spent most resources on a "war on drugs" targeting drug supply and incarcerating users often for long periods and with devastating societal, personal and family consequences. Policies appear more often than not to be guided by societal values and commercial pressures with scientific data given a back seat. There is clear data to suggest that targeting help for the abusers themselves will be more productive and less costly than other policies in the long-term. In the words of a review from RAND *How Goes the War on Drugs*—" In contrast to enforcement, treatment can cost-effectively reduce the consumption of an endemic drug and thus also reduce the harms associated with use" (Caulkins et al. 2005).

One key observation is that age of drug onset is a strong predictor for continued abuse in adulthood suggesting that drug exposure to a young brain has considerably more impact on subsequent behavior than exposing an older brain. This is quite logical given the increase in learning ability of young versus old brains. Statistics of drug use in under 18's are particularly worrisome in this regard particularly the estimate that, in recent years, about 3% of 8th graders (13–14 years old) and 10% of 12th-graders (17–18 years old) illicitly took Vicodin or another opioid pharmaceutical during the prior year. This suggests that as a matter of urgency, we need to focus drug prevention resources on the education of the brain at a young age to tame and direct the reward neuronal engine to constructive endpoints.

Conclusions

The statistics across the US and indeed the world paint a bleak picture of our attempts to reconcile the operation of the neuronal circuitry in the brain regulating reward and hedonic tone within the context of current day society. One could speculate that, because society now readily supplies the natural rewards the system was designed to pursue, the reward system and the executive or goal-directed action system are today largely superfluous and the control of actions switched largely to habit circuits leaving an abandoned goal-directed system seeking objects of stimulation. It is clear that education of the reward system with drugs can be disastrous, since it results in a dysphoric state in absence of the drug and more rapid and extensive control by habit processes—the mantra has been just say no, but what are the alternatives? How does one educate or re-educate a population to have a healthy hedonic tone and reward system and how much is inevitable from the roll of the genetic dice? These are challenges for our future.

References

Alberini, C. M., Milekic, M. H., & Tronel, S. (2006). Mechanisms of memory stabilization and de-stabilization. *Cellular and Molecular Life Sciences, 63,* 999–1008.

Aron, J. L., & Paulus, M. P. (2007). Location, location: Using functional magnetic resonance imaging to pinpoint brain differences relevant to stimulant use. *Addiction 102*(Supp), 33–43.

Balleine, B. W. (2005). Neural bases of food-seeking: Affect, arousal and reward in corticostriatolimbic circuits. *Physiology & Behavior, 86,* 710–717.

Balleine, B. W., Delgado, M. R., & Hikosaka, O. (2007). The role of the dorsal striatum in reward and decision-making. *Journal of Neuroscience 27,* 8161–8165.

Bryant, C. D., Zaki, P. A., Carroll, F. I., & Evans, C. J. (2005). Opioids and addiction: Emerging pharmaceutical strategies for reducing reward and opponent processes. *Clinical Neuroscience Research, 5*(1), 03–115.

Caulkins, J. P., Reuter, P., Iguchi, M. Y., & Chiesa, J. (2005). *How goes the "War on Drugs"?*. Santa Monica: RAND Corporation.

Gaveriaux-Ruff, C., & Kieffer, B. L. (2002). Opioid receptor genes inactivated in mice: The highlights. *Neuropeptides, 36,* 62–71.

Hiroi, N., & Agatsuma, S. (2005). Genetic susceptibility to substance dependence. *Mol Psychiatry, 10,* 336–344.

Koob, G. F. (1992). Neural mechanisms of drug reinforcement. *Annals of the New York Academy of Sciences, 654,* 171–191.

Leshner, A. I. (1997). Addiction is a brain disease, and it matters. *Science, 278,* 45–47.

Miller, M., Vorel, S. R., Tranguch, A. J., Kenny, E.T., Mazzoni, P., van Gorp, W. G., & Kleber, H. D. (2006). Anhedonia after a selective bilateral lesion of the globus pallidus. *The American Journal of Psychiatry 163,* 786–788.

Milner, P. M. (1991). Brain-stimulation reward: A review. *Canadian Journal of Psychology, 45,* 1–36.

Pecina, S., Smith, K. S., & Berridge, K. C. (2006). Hedonic hot spots in the brain. *Neuroscientist,* 500–511.

Schafer, W. R. (2004). Addiction research in a simple animal model: The nematode Caenorhabditis elegans. *Neuropharmacology 47*(Suppl 1), 123–131.

Skoubis, P. D., & Maidment, N. T. (2003). Blockade of ventral pallidal opioid receptors induces a conditioned place aversion and attenuates acquisition of cocaine place preference in the rat. *Neuroscience, 119,* 241–249.

Skoubis, P. D., Lam, H. A., Shoblock, J., Narayanan, S., & Maidment, N. T. (2005). Endogenous enkephalins, not endorphins, modulate basal hedonic state in mice. *European Journal of Neuroscience, 21,* 1379–1384.

Wise, R. A. (2004). Dopamine, learning and motivation. *Nature Reviews Neuroscience, 5,* 483–494.

Brain Research: Improving Social Harmony on Reward, Trust, and Impulse Control

Theodor Landis

Global harmony implies harmonic social interaction between humans. Just looking at everyday TV news makes it clear that neither on the small (individuals) nor on the large scale (nations) humans manage harmonious interactions. Are we made to quarrel and to kill each other usually in affectively stressing situations leading to loss of control? Might brain research uncover some inbuilt mechanisms switching us from caring fathers or mothers to murderous fools?

As a representative of the Swiss National Science Foundation (SNF) I will talk about some key brain research experiments supported by the SNF which have led to some insight in mechanisms of human social interaction. The three topics I will briefly touch arc "reward" "trust" and "impulse control".

In the mid nineties the neurophysiologist Wolfram Schultz and his group at the University of Fribourg found dopamine neurons in the monkey's midbrain to code for the prediction and expectation of reward. Their work has triggered a tremendous amount of brain research on reward behavior in animals and humans during the last 20 years. Reward, be it social, monetary or affective, is what we are seeking for in most of our actions. It provides also a link to the other theme of this conference, education, since reward is one of the most powerful forces of learning.

In 2002 the economist Ernst Fehr and his group at the University of Zurich published an article on "altruistic punishment in humans". Coming from a completely different research field, economics, they introduced the notion of "trust" into brain research and provided a methodology to test tor "trust". They and others have subsequently identified brain structures concerned with the modulation of trust. Trust is certainly by far the most important human quality to guarantee harmony in human social interaction. Brain research on trust is at its beginning but it has a potential to increase "social harmony".

T. Landis (✉)
University of Geneva, Geneva, Switzerland
e-mail: Theodor.Landis@unige.ch

Impulse control disorders, such as addictions, have long been known and studied in psychiatry and neurology. When due to brain damage they are usually related to frontal lobe structures. In the last ten years risk-taking behavior, particularly in gambling tasks, has extensively been investigated. Very recently Daria Knoch and colleagues at the University of Zurich could demonstrate that repetitive transcranial magnetic stimulation over the right, but not the left frontal lobe in normal humans could not only influence risk-taking, but also reciprocal fairness. Impaired control of our impulses is a common burden of harmonic social behavior and the finding that this control can be externally modified may eventually open therapeutic avenues to improve global harmony.

Introduction

I am very honored to speak in front of such a distinguished audience. I have been invited as a representative and member of the executive committee of the Swiss National Research Science Foundation (SNF), the main granting office of my country. They have chosen me because I am the only behavioral neurologist and brain researcher in this team.

Switzerland is a very tiny country without any natural resources, but it has a long tradition in supporting basic science. A recent analysis on the "scientific impact of nations" published in Nature (1) comparing among other things "wealth intensity" with "citation intensity" gives us the credit of efficiently supporting science with respect to our wealth.

But does this scientific achievement really improve social interaction or global harmony?

The basis of the way humans socially interact lies in our brains. But does brain research address these questions? And might it provide ultimately a tool to improve global harmony?

Brain research has as other sciences undergone cycles. While early stimulation experiments of deep brain structures in cats led to full-blown behavior actions such rage, mating behavior and skep, and led to a Swiss Nobel prize in 1949 to W.R. Hess, the advent of the possibility of single cell recordings shifted the research questions away from behavior towards primary and associative processing of the senses, in particular of vision.

It is only relatively recently that neurophysiologists have rediscovered an interest to study emotions, motivations, reward seeking, risk-taking, impulse control and social interactions.

In an attempt to combine the theme of this conference "brain research: improving social harmony" with my being invited as representative of the Swiss National Science Foundation I will talk about some key experiments in brain research supported by the SNF, which have led to some insight in mechanisms of human motivation and social interaction, and which have triggered more brain research, hopefully ultimately leading to "improving social harmony". The three topics I will briefly touch are "reward", "trust" and "impulse control".

Reward

Behavioral experiments suggest that learning is driven by changes in the expectations about future salient events such as rewards. In conditioning experiments a sensory cue, for example a light, when consistently associated with a reward, will induce a prediction about the likely time and magnitude of the reward. If the light fully predicts the reward, the animal will not learn to make an association to any additional sensory cue, for example a sound, since it adds no further predictive value. This suggests that reward-dependent learning is driven by the unpredictability of the reward by the sensory cue, that is by the deviation or "errors" between the predicted time and magnitude of reward and their actual experienced times and magnitudes. Knowing more about reward-associated learning is essential for a better understanding of learning per se, motivation, but also pathological states of reward seeking.

In a series of experiments during the last 20 years (2), in which activity of single dopamine neurons was recorded in the midbrain ventral tegmental area (VTA) of alert monkeys, the neurophysiologist Wolfram Schultz and his group at 'the University of Fribourg/Switzerland linked the above mentioned learning theory on one hand to a specific neurotransmitter system, the dopamine system, on the other hand to a small group of well-defined nerve cells, the dopamine neuron of the VTA, and moreover, to an algorithm, the temporal difference method, widely used in engineering. They discovered a dynamic in the output coding of the VTA dopamine neurons in relation to the prediction error. First, they found that dopamine neurons respond with short, phasic activations when presented with unpredicted appetitive stimuli such as fruit juice—hence a positive error in the prediction of reward; second, they showed that once a sensory cue was learned and fully predicted the occurrence of the reward—hence no prediction error, the dopamine neurons would be activated by the reward predicting sensory cue, but failed to be activated by the predicted reward, and third, they showed that when the predicted reward fails to occur the response of the neuron is depressed exactly at the time the reward would have occurred. They thus discovered a brain circuitry with the capacity of self-organization of behavior with prediction errors, optimizing magnitude, probability and time of reward. Their work has triggered a tremendous amount of brain research on reward behavior in animals and humans during the last 20 years. Reward, be it social, monetary or affective, is what we are seeking for in most of our actions. It provides also a link to the other theme of this conference, education, since reward is one of the most powerful forces of learning. I will not go into the long underestimated importance this dopamine system has in regulating learning, addiction and impulse control (see Evans, this volume).

Trust, Fairness and Altruistic Punishment

Social harmony will be reached when all humans cooperate harmoniously, a dream of all of us, however, rarely achieved. In order to address the question of social harmony in an attempt to improve it, we imperatively will have to study human cooperation and possibly find its cerebral basis.

Human cooperation is an evolutionary puzzle. Unlike any other species, humans frequently cooperate with genetically completely unrelated strangers, with large groups of individuals, and with peoples they will never meet again. Big game hunting, warfare or conserving common property resources throughout times necessitated cooperation between humans and "the public good" profited to every member of the group, even to those who did not pay any costs of providing the good. However, there is the problem of those who free ride on the cooperation of others. Punishment of such free riders could solve the problem, but who would be willing to punish free riders on his own costs without material benefits? If humans had an inherent tendency for such "altruistic punishment" enduring cooperation could be "forced" and pay off.

This question was addressed by the economist Ernst Fehr and his group (3) at the University of Zurich in a "public goods" experiment with a large number of students playing, by various combinations of 4 participants via computer communication and real money gains, games with and without punishment conditions. Each participant received 20 monetary units (MU) of which he could invest as much as he wanted into a group project. Irrespective of how much he invested he would receive 0.4 MU per MU invested in the group project by all four participants. Thus, if no participant would invest in the group project each one would keep 20 MUs, while if all invested their 20 MUs each participant would earn 32 MUs (0.4 × 80). Participants remained anonymous and investment but also punishment decisions were made simultaneously. Once the investment decision was made, participants were informed about the investment of the other group member. Once informed, the participants could in the punishment condition, punish each group member by assigning 0–10 points. Each point would cost the punished member 3 MUs and the punisher 1 MU.

Punishment was very frequent and followed a clear pattern. Below-average contributors were punished by cooperators. This "altruistic punishment" of non-cooperators substantially increased the amount that subjects invested in the public good. In the non-punishment condition the initial mean group investment was significantly lower than in the punishment condition and decreased with every new game to a 5/20 MU, while in the punishment condition it significantly increased with every game to a 16/20 MU. Subjective emotional ratings of anger towards non-cooperators followed a similar pattern witnessing a clear awareness of both cooperators and non-cooperators about the emotional impact of their non-fairness. This study has profound implications for the understanding of human cooperation, which probably is the basis of social harmony. It suggests that "altruistic punishment", driven by an emotional reaction to unfairness may be the motor to make cooperation work. Moreover, it suggests that this mechanism may be implemented in the human brain.

In a follow-up experiment the same group of "neuro-economists" (4) attempted to uncover the neural basis of altruistic punishment, using a design based on "trust" and intentional abuse of trust. Two anonymous players A and B, both endowed with 10 MUs can substantially increase their income if they trust each other and act in a trustworthy manner. Every MU that A sends to B is quadrupled. If he sends all 10 MUs, B receives 40 MUs. If B is fair, he sends back 25 MUs, if he is not trustworthy he sends nothing back.

However, A receives the option of punishing B with up to 20 punishment points. There are four conditions: (1) B abuses intentionally and punishment is costly, (2) B abuses intentionally and punishment is free, (3) B abuses intentionally and punishment is symbolic, and (4) D abuses non-intentionally (random generated) and punishment is costly. The experiment took place in a PET scanner and the one minute between the unfair return and the punishment decision was scanned. Since there was no opportunity to really punish in the symbolic condition and no desire to punish in the non–intentional condition, these two conditions were compared to those in which punishment was effective and provided some satisfaction to the punisher. Effective punishment activated the caudate nucleus, a structure which has been implicated in making decisions or taking actions that are motivated by anticipated reward. Moreover, the more the nucleus caudatus was activated, the more subjects were willing to spend money to punish. Our brains thus seem to reward punishment of unfair behavior or deceived trust.

However, the reward seeking desire to punish had to be weighted against the costs. While most people punished maximally when punishment was free, punishment was significantly reduced when it incurred costs. The comparison of these two conditions activated the ventromedial prefrontal and the medial orbitofrontal cortex structures involved in complex decision making and choices that require the coding of reward value. Thus punishing is not automatic but the satisfaction reward desire is controlled by "higher-order" integration areas.

Impulse Control, Reciprocal Fairness and Its Modification

The ability to make correct decisions in a complex and changing environment requires careful weighting of risks and benefits. As mentioned above the prefrontal cortex appears to be critical in such decision making processes. When these structures are damaged impulse control disorders in the form of risk-taking or addictive behavior can occur. When due to focal brain damage it frequently occurs subsequent to right frontal lesions. Impulse control disorders such as pathological gambling or substance abuse are usually detrimental to the individual and the social harmony of their environment. However, this is not always the case. We recently described an addictive compulsion towards "fine eating" the gourmand syndrome, which corresponds to all the criteria of an impulse control disorder, is strongly related to right anterior brain lesions, but is benign.

Not only fundamental brain research, but also recent clinical observations relate impulse control disorders to the dopamine reward system. Patients with Parkinson's disease, a degenerative disorder of midbrain dopaminergic neurons, may, when overdosed with L-dopa, or over-stimulated with deep brain stimulation develop transient impulse control failure. They may transitorily become pathological gamblers, compulsive shoppers, compulsive eaters, hypersexuals or even compulsive singers.

In the last 10 years risk-taking behavior, particularly in gambling tasks, has extensively been investigated and been associated with frontal lobe structural damage. In the last 20 years repetitive transcranial magnetic stimulation (rTMS), a new tool to modify brain function in humans, has been developed. Depending upon the frequency of stimulation rTMS can transiently inhibit or excite the neuronal function or the underlying cortex.

Quite recently, Daria Knoch and the neuropsychology group (5) at the University of Zurich used this technique to study risk-taking behavior in normal subjects. They disrupted left or right dorsolateral prefrontal cortex function with low-frequency rTMS before applying a gambling paradigm. This Risk Task which varies winning probabilities with rewards and penalties measures decision making under risk. They found that people showed a significantly riskier decision-making after the disruption of function of the right, but not the left prefrontal cortex, suggesting that the right prefrontal lobe plays a crucial role in the suppression of superficially seductive options. Besides confirming the functional asymmetry of the prefrontal lobes in decision-making it shows that this fundamental human capacity can be, though transitorily, manipulated by cortical stimulation. The potential to modify risk-taking behavior may in the future offer therapeutic options for impulse control disorders such as substance abuse or pathological gambling.

Could functional modification by rTMS also he used to modify social interactions? Daria Knoch and the Zurich neuroeconomy group (6) tested this question by using the "ultimate game" under the condition of left or right dorsolateral prefrontal lobe rTMS stimulation. The "ultimate game" illustrates the tension between economic self-interest, and reciprocity and fairness. In the "ultimate game" a proposer and a receiver have to agree on the division of a given amount of money. The proposer makes a single suggestion of how the amount is divided and the receiver can either accept or reject it. If he rejects the offer, both players will earn nothing. If he accepts, he will earn what is proposed. If purely economic self-interest would drive the receiver's decision he would have to accept even very low offers, since earning little is better than earning nothing. However, if concerns for reciprocity drive his decisions, he will reject unequal and thus unfair offers. The receiver will be torn between considerations of economic self-interest and concerns about reciprocal fairness.

Disruption of the function of the right, but not the left dorsolateral prefrontal cortex by rTMS showed a significant increase in the acceptance of unfair offers.

Thus a switch towards economic self-interest, away from fairness considerations takes place when the functions of the right prefrontal cortex are inhibited. This structure thus seems necessary to maintain a control of human selfishness in favor of "social harmony". Interestingly enough this right prefrontal stimulation did only interfere with the actual acceptance of unfair offers, but not with the individuals judgment of the degree of fairness of the offer which remained intact.

The implications of this study to social harmony are manifold: they show the right prefrontal lobe to be instrumental in controlling the conflict induced by the divergence of self-interest and fairness motives. They also show that this controlling instance is externally modifiable, thus potentially accessible to therapies enhancing fairness and social harmony.

Conclusions

Brain research has during the past 20 years made tremendous progress in understanding the basis of social interaction and cooperation. I have shown only a few key experiments chosen because they are supported by the Swiss National Science Foundation for which I speak here as a representative, but the experimental effort in this domain has grown exponentially. Brain research has discovered a whole neuronal system in the brain dedicated to human social behavior.

This system comprises phylogenetically ancient areas in the midbrain concerned with reward predictions, striatal areas providing reward for punishing those who do not behave, those who are selfish and not concerned with public goods, up to phylogenetically new areas in the ventromedial and dorsolateral area of the frontal brain which weighs the advantages of behaving in favor of society or of the self. However, our knowledge of the wiring, the computational algorithms and the biochemistry of this "social brain" are still very incomplete and immature and far from providing "practicable" therapeutic options to improve "social harmony".

I have been asked to provide some ideas about "actions to take". To take actions on the basis of incomplete knowledge is one of the common pitfalls in therapeutic medicine. Not completely understood pathophysiological mechanisms of disease has, in the past, led to disastrous treatment attempts. On the basis of the above mentioned experiments it would seem, that enhancing, via teaching and publicity, the acceptance of "altruistic punishment", could be a practical and physiological approach to enhancing human social cooperation. But is our knowledge about these implemented brain mechanisms complete enough to avoid negative drawbacks?

One action to take is certainly to support brain resear.ch in this domain to complete our knowledge about the "social brain".

References

de Quervain, D., Fischbacher, U., Treyer, V., Treyer, V., Schellhammer, M., Schnyder, U., et al. (2004). The neural basis of altruistic punishment. *Science, 305,* 1254–1258.

Fehr, E., & Glichter, S. (2002). Alttuistic punishment in humans. *Nature, 415,* 137–140.

King, D. A. (2004). The scientific impact of nations. *Nature 430,* 311–3 16.

Knoch, D., Gianolli, L. R. R., Pascual-Leone, A., Treyer, V., Regard, M., Hohmann, M., et al. (2006a). Disruption of right prefrontal cortex by low -frequency repetitive transcranial magnetic stimulation induces risk-taking behavior. *Journal of Neuroscience, 26,* 6469–6472.

Knoch, D., Pascual-Leone, A., Meyer, K., Treyer, V., & Fehr, E. (2006b). Diminishing reciprocal fairness by disrupting right prefrontal cortex. *Science, 314,* 829–832.

Schultz, W., Dayan, P., & Montague, P. R. (1997). A neural substrate of prediction and reward. *Science 275,* 1593–1598.

Theodor Landis is honorary professor of neurology at the University of Geneva . He presented this paper as a representative of the Swiss National Science Foundation.

Making Sense of Neurodegeneration: A Unifying Hypothesis

Barry Halliwell

The neurodegenerative diseases that afflict humans have different origins, affect different parts of the nervous system and show different pathologies and prognoses. Much current research into such diseases emphasizes genetic risk factors. Indeed, Huntington's disease is purely genetic-if you are unlucky enough to have the abnormal gene on your chromosome 4, you arc fated to get the disease: and (as far as we know), diet and lifestyle cannot influence this (Ramaswamy et al. 2007). By contrast, almost all cases of Parkinson's disease (PD) seem not to involve defective genes, and it is widely thought that one or more environmental toxins cause this "shaking palsy" (Hardy et al. 2006; Moore et al. 2005; Halliwell 2006a). Only about 10% of cases of motor neuron diseases have a genetic origin (Mitchell and Borasio 2007). Similarly, genetically determined cases of Alzheimer's disease (familial AD) are rare, although genetic predispositions are common [61]. One involves the gene encoding apoprotein E, on chromosome 19. This has three alleles. The most common is *apoE3,* which encodes a protein with a cysteine residue at position 112. In *apoE2* the protein encoded has an extra cysteine at 158 (replacing arginine) whereas *apoE4* encodes a protein in which cys112 is replaced by arg. Possession of the *apoE4* allele is a significant risk factor for the development of AD, whereas *opoE2* seems to lower the risk (Selkoe 2005; Chai 2007).

B. Halliwell (✉)

Department of Biochemistry and Office of the President National University of Singapore, University Hall Lee Kong Chian Wing, UHL #05-02G 21 Lower Kent Ridge Road, Singapore 119077, Singapore

e-mail: bchbh@nus.edu.sg

© Springer International Publishing AG, part of Springer Nature 2019

J. A. S. Kelso (ed.), *Learning To Live Together: Promoting Social Harmony,*

https://doi.org/10.1007/978-3-319-90659-1_13

What Does My Laboratory Do?

My laboratory has for many years researched the importance of free radicals, related "reactive species" and antioxidants in human health and disease (Halliwell and Gutteridge 2007). Thus we were naturally drawn to study neurodegeneration because all the major neurodegenerative diseases show increased levels of damage by reactive species ("oxidative damage") in the affected brain regions (Halliwell and Gutteridge 2007; Halliwell 2006b; Butterfield et al. 2007; Przedborski and Ischiropoulos 2005).

We are also interested in what else neurodegenerative diseases have in common—this includes impaired mitochondrial function, the presence of aggregated proteins, alterations in iron metabolism (usually increased iron deposition in the affected brain regions) and some involvement of inflammation and of excitotoxicity. As a result of our studies and those by many others, we have advanced the concept (Halliwell 2006a, b) that neurodegeneration, particularly in PD, can be triggered in at least 3 ways:

- agents entering the brain that cause prolonged oxidative stress (defined as increased free radical formation beyond the capacity of antioxidant defense systems to cope (Halliwell and Gutteridge 2007)),
- agents entering the brain that inhibit mitochondrial function (the PD-inducing neurotoxin MPTP is known to act in this way (Fukae et al. 2007)).
- agents entering the brain that interfere with protein turnover (removal of unwanted normal proteins, and of abnormal proteins) by the proteasome system. Once triggered, all these events interplay in a complex manner.

Let us consider an example. In some studies, treatment of rats or monkeys with low doses of rotenone, an inhibitor of mitochondrial function, over long periods produces PD-like symptoms and neurodegeneration accompanied by oxidative damage and generation of inclusion bodies (protein aggregates) (Przedborski and Ischiropoulos 2005). Unlike MPP + (the active metabolite of MPTP mentioned above (Fukae et al. 2007)), rotenone does not concentrate in dopamine neurons, yet it can still induce fairly selective neurodegeneration in the substantia nigra (SN), the main affected region in PD. It follows that SN neurons may, for some reason, be especially sensitive to complex I inhibition, so that any toxin affecting complex 1 might cause PD-like neurodegeneration (Halliwell 2006b; Caboni et al. 2004; Champy et al. 2004). Such toxins may be widespread in the environment; even rotenone in some places (Przedborski and Ischiropoulos 2005; Fukae et al. 2007; Caboni et al. 2004; Champy et al. 2004; Betarbet et al. 2006). So might proteasome inhibitors; many are natural products (Bogyo and Wang 2002). The phenolic "antioxidants" BO-653 and probucol were reported to decrease the gene expression and levels of the proteasome in human endothelial cells (Takabe et al. 2001), suggesting that many more agents than we currently suspect may modulate proteasome function. Indeed, when we screened a range of medicinal plants, many contained compounds able to inhibit one or more aspects or proteasome function (data in preparation for publication).

However, PD need not always start with mitochondrial defects. Studies with 6-hydroxydopamine, an agent that oxidizes rapidly to cause oxidative stress, reveal that oxidative stress can cause neurodegeneration (Przedborski and Ischiropoulos 2005). Dopamine oxidation products (which accumulate in PD) can both damage mitochondria (this damage itself leading to increased free radical production) and inactivate the proteasome, causing unwanted proteins to accumulate and aggregate. Mitochondrial inhibitors, proteasome inhibitors and agents that induce oxidative stress may be widespread in nature, could possibly act synergistically, and industrial activities may introduce more of them (Halliwell 2006a, b; Halliwell and Gutteridge 2007; Sun et al. 2005; Canesi et al. 2003; Frigerio et al. 2006).

Considering Other Factors

As we have seen, genetic predispositions alter the risk of developing AD. But they are not the whole story, nor possibly even a major part of it. Other risk factors include a low dietary intake of folic acid (McCaddon 2006; Snowdon et al. 2000), high blood cholesterol levels (Lahiri et al. 2007; Ong and Halliwell 2004), high plasma homocysteine (McCaddon 2006), repeated minor brain trauma (Schmidt et al. 2001) , and even a low level of education ("use the brain or lose it"), as dramatically illustrated by the Nun study (Snowdon et al. 2000; Snowdon 2001; Riley et al. 2005). The Nun study examined 678 Catholic Sisters, living under defined conditions in convents. They agreed to undertake annual medical examinations and cognitive testing, and some were followed over more than 30 years. Many agreed to donate their brains for autopsy after death, from which it was discovered that pathology does not always equate with clinical status. The hallmark pathology of AD is the presence of a large number of senile plaques and neurofibrillary tangles in the brain (Selkoe 2005; Chai 2007; Riley et al. 2005). Yet the Nun study found cases of nuns completely *compos mentis* in life with marked pathology after death, i.e. an "AD brain", as identified pathologically, can co-exist with normal cognitive function (Snowdon 2001; Riley et al. 2005). Other factors must therefore contribute to the appearance of clinical dementia. These can be co-existing other diseases, e.g. small strokes (Snowdon 2001) or vascular impairment due to atherosclerosis. Or perhaps a high level of education/intelligence allows compensation for significant neuronal dysfunction and loss. In the Nun study, the four major factors that influenced risk of developing AD were identified as *apoE* gene status, folic acid intake, level of education and the essay that each nun wrote on entering the convent-the greater the richness of the language, the lower the risk of clinical dementia (Snowdon 2001; Riley et al. 2005).

Other aspects of diet, and exercise, may also play a role. Indeed, several studies have indicated that exercise decreases the risk of dementia and slows the progression of cognitive decline in AD patients (Podewils et al. 2005; Arkin 2007;

Briones 2006). Studies in mice suggest at least one mechanism-that of decreasing oxidative damage. Levels of free radical-damaged proteins in the brains of old mice were decreased after a program of regular exercise (Radak et al. 2006). Other studies suggest that diets rich in polyunsaturated fatty acids such as docosahex-aenoic acid may be beneficial in slowing AD development (Connor and Sonnor 2007). Animal studies have also detected beneficial effects of feeding polyphenol-rich foods such as blueberries (Lau et al. 2007), and of restrictions in food intake (Malison et al. 2004). Whether polyphenols get into the human brain is uncertain as yet, however.

All the factors mentioned above are modifiable risk factors-one can control folic acid intake, plasma homocysteine, plasma cholesterol, exercise levels and diet (PUFAs, vitamins, calories, fruit/vegetable intake). Given the fast growing elderly population and the strong age-related nature of AD and other neurodegenerative diseases, much more work needs to be done on the relation of diet and behavior to neurodegeneration. Smoking cigarettes decreases the risk of developing PD (Ritz et al. 2007)-not a behavior one wishes to encourage because smoking generates many other ills-but perhaps a clue to trigger further research into the origins of the disease.

Conclusions

1. We need to mount a search for environmental neurotoxins (natural and industrial chemicals) that can enter the brain-there may be more than we think.
2. We need to understand the mechanism and significance of the nutritional and other factors that can influence the onset and progression of neurodegeneration, and implement these changes in lifestyle, especially in patients with genetic predispositions to neurodegenerative diseases.

References

Arkin, S. (2007). Language-enriched exercise plus socialization slows cognitive decline in Alzheimer's disease. *Journal of Alzheimer's Disease and Other Dementias, 22*(1), 62–77.

Betarbet, R., Canet-Aviles, R. M., Sherer, T. B., Mastroberardino, P. O., McLendon, C., Kim, J. H., Lund, S., Na, H. M., Taylor, G., Bence, N. F., Kopito, R., Seo, B. B., Vagi, T., Vagi, A., Klinefelter, G., Cookson, M. R., & Greenamyre, J. T. (2006). Intersecting pathways to neurodegeneration in Parkinson's disease: Effects of the pesticide rotenone on DJ-I, alpha-synuclein, and the ubiquitin-proteasome system. *Neurobiology of Disease, 22*(2), 404–420.

Bogyo, M., & Wang, E. W. (2002). Proteasome inhibitors: complex tools for a complex enzyme. *Current Topics in Microbiology and Immunology, 268*, 185–208.

Briones, T. L. (2006). Environmental, physical activity, and neurogenesis: implications for prevention and treatment of Alzheimer's disease. *Current Alzheimer Research, 3*(1), 49–54.

Butterfield, D. A., Reed, T., Newman, S. F., & Sultana, R. (2007). Roles of amyloid beta-peptide-associated oxidative stress and brain protein modifications in the pathogenesis of Alzheimer's disease and mild cognitive impairment. *Free Radical Biology and Medicine, 43* (5), 658–677.

Caboni, P., Sherer, T. B., Zhang, N., Taylor, G., Na, H. M., Greenamyre, J. Y., et al. (2004). Rotenone, deguelin, their metabolites, and the rat model of Parkinson's disease. *Chemical research in Toxicology, 17*, 1540–1548.

Canesi, M., Perbellini, L., Maestri, L., Silvani, A., Zecca, L., Bet, L., et al. (2003). Poor metabolization of n-hexane in Parkinson's disease. *Journal of Neuralogy, 250*(5), 556–560.

Chai, C. K. (2007). The genetics of Alzheimer's disease. *American Journal of Alzheimer's Disease and Other Dementias, 22*(1), 37–41.

Champy, P., Hoglinger, G. U., Feger, J., Gleye, C., Hoequemiller, R., Laurens, A., Guerineau, V., Laprevote, O., Medja, F., Lombes, A., Michel, P. P., Lannuzel, A., Hirch, E. C., & Rubery, M. (2004). Annonacin, a lipophilic inhibitor of mitochondrial complex I, induces nigral and striatal neurodegeneration in rats: possible relevance for atypical parkinsonism in Guadeloupe. *Journal of Neurochemistry, 88*(1), 63–69.

Connor, W. E., & Sonnor, S. J. (2007). The importance of fish and docosahexaenoic acid in Alzheimer disease. *American Journal of Clinical Nutrition, 85*, 929–930.

Frigerio, R., Sanft, K. R., Grossardt, S. R., Peterson, B. J., Elbaz, A., Bower, J. H., Ahlskog, A. E., de Andrade, M., Maraganore, O. M., & Rocca, W. A. (2006). Chemical exposures and Parkinson's disease: A population-based case-control study. *Movement Disorders, 21*(10), 1688–1692.

Fukae, J., Mizuno, T., & Hattori, N. (2007). Mitochondrial dysfunction in Parkinson's disease. *Mitochondrion, 7*(1–2), 58-62.

Halliwell, B. (2006a). Proteasomal dysfunction: A common feature of neurodegenerative diseases? Implications for the environmental origins of neurodegeneration. *Antioxidants and Redox Signaling, 8*, 2007–2019.

Halliwell, B. (2006b). Oxidative stress and neurodegeneration: Where are we now? *Journal of Neurochemistry, 97*, 1634–1658.

Halliwell, B. & Gutteridge, J. M. C. (2007). *Free radicals in biology and medicine* (4th ed.). Clarendon Press, Oxford, UK.

Hardy, J., Cai, H., Cookson, M. R., Gwin-Hardy, K., & Singleton, A. (2006). Genetics of Parkinson's disease and parkinsonism. *Annals of Neurology, 60*(4), 389–398.

Lahiri, D. K., Maloney, S., Basha, M. R., Oe, Y. W., & Zawia, N. H. (2007). How and when environmental agents and dietary factors affect the course of Alzheimer's disease: the "Learn" mood (latent early-life associated regulation) may explain the triggering of AD. *Current Alzheimer Research, 4*(2), 219–229.

Lau, F. C., Shukitt-Hale, B., & Joseph, J. A. (2007). Nutritional intervention in brain aging: reducing the effects of inflammation and oxidative stress. *SubCellular Biochemistry, 42*, 299–318.

Malison, M. P., Duan, W., Wan, R., & Guo, Z. (2004). Prophylactic activation of neuroprotective stress response pathways by dietary and behavioral manipulations. *NeuroRx, 1*, 111–116.

McCaddon, A. (2006). Homocysteine and cognition-a historical perspective. *Journal of Alzheimers Disease, 9*(4), 361–380.

Mitchell, J. D., & Borasio, G. D. (2007). Amyotrophic lateral sclerosis. *Lancet, 369*(9578), 2031–2041.

Moore, D. J., West, A. B., Dawson, V. L., & Dawson, T. M. (2005). Molecular pathophysiology of Parkinson's disease. *Annual Review of Neuroscience, 28*, 57–87.

Ong, W. Y., & Halliwell, B. (2004). Iron, atherosclerosis, and neurodegeneration: A key role for cholesterol in promoting iron-dependent oxidative damage? *Annals of the New York Academy of Sciences, 1012,* 51–64.

Podewils, L. J., Guallar, E., Kuller, L. H., Fried, L. P., Lopez, O. L., Carlson, M., et al. (2005). Physical activity, APOE genotype, and dementia risk: findings from the Cardiovascular Health Cognition Study. *American Journal of Epidemiology, 161,* 639–651.

Przedborski, S., & Ischiropoulos, H. (2005). Reactive oxygen and nitrogen species: Weapons of neuronal destruction in models of Parkinson's disease. *Antioxidants and Redox Signaling, 7,* 685–693.

Radak, Z., Toldy, A., Szabo, Z., Siamilis, S., Nyakas, C., Silye, G., et al. (2006). The effects of training and detraining on memory, neurotrophins and oxidative stress markers in rat brain. *Neurochemisty International, 49*(4), 387–392.

Ramaswamy, S., Shannon, K. M., & Kordower, J. H. (2007). Huntington's disease: pathological mechanisms and therapeutic strategies. *Cell Transplantation, 16*(3), 301–312.

Riley, K. P., Snowdon, D. A., Desrosiers, M. F., & Markesbery, W. R. (2005). Early life linguistic ability, late life cognitive function, and neuropathology: findings from the Nun study. *Neurobiology of aging, 26*(3), 341–347.

Ritz, B., Ascherio, A., Checkoway, H., Marder, K. S., Nelson, L. M., Rocca, W. A., Ross, O. W., Strickland, D., Van Den Eeden, S. K., Gorell, J. (2007). Pooled analysis of tobacco use and risk of Parkinson disease. *Archives of Neurology, 64*(7), 990–997.

Schmidt, M. L., Zhukareva, C., Newell, K. L., Lee, V. M., & Trojanowski, J. Q. (2001). Tau isoform profile and phosphorylation state in dementia pugilistica recap itulate Alzheimer's disease. *Acta neuropathologica (Berl), 101*(5), 518–524.

Selkoe, O. J. (2005). Defining molecular targets to prevent Alzheimer disease. *Archives of Neurology, 62*(2), 192–195.

Snowdon, D. (2001). *Aging with grace: What the nun study teaches us about leading, longer, healthier, and more meaningful lives.* New York: Bantam Books.

Snowdon, O. A., Tully, C. L., Smith, C. D., Riley, K. P., & Markesbery, W. R. (2000). Serum folate and the severity of atrophy of the neocortex in Alzheimer disease: Findings from the Nun study. *The American Journal of Clinical Nutrition lV'lllr, 71*(4), 993–998.

Sun, F., Ananthararn, V., Latchownycandane, C., Kanthasamy, A., & Kanthasamy, A. G. (2005). Dieldrin induces ubiquitin-proteasorne dysfunction in alpha-synuclein overexpressing dopaminergic neuronal cells and enhances susceptibility to apoptotic cell death. *Journal of Pharmacology and Experimental Therapeutics, 315,* 69–79.

Takabe, W., Kodama, T., Hamakubo, T., Tanaka, K., Suzuki, T., Aburatani, H., Matsukawa, K., & Noguchi, N. (2001). Anti-atherogenic antioxidants regulate the expression and function of proteasome alpha-type subunits in human endothelial cells. *Journal of Biological Chemistry, 276*(44), 40497-40501.

The Brain as a Learner/Inquirer/ Creator: Some Implications of Its Organization for Individual and Social Well Being

Paul Grobstein

At its best, empirical research opens new and fruitful directions for exploring what it means to be human. In this essay, I will describe several aspects of research on the brain that seem to me to have this character, suggesting that we can and should think of ourselves, individually and collectively, as active creators and revisers of meaning, for ourselves and for the world we find ourselves in. I will conclude with several social policy recommendations that follow from this perspective.

The past 50 years have seen explosive growth in human understandings of ourselves as biological entities, not only in narrow terms but in quite broad ones as well. Empirical science has given us new appreciation not only of mechanisms of inheritance but also of the complex processes by which genetic information influences all aspects of humanness, and of the evolutionary process that creates that genetic information and links us with each other and with other living organisms in an ongoing exploration of the possibilities of life. We have as well new understandings of the brain, not only of the details of its internal workings in relation to particular behaviors but also of how its complexity yields rich potentials for the continuing development and evolution of humanness, both individually and collectively. In this essay, I want to focus on several of these understandings that seem to me of particular importance for the future, for creating new conceptions of humanness that could in turn yield richer and fuller lives for all human beings.

An Historical Context: Genes, Behaviorism, and the Brain

In 1971 two books were published, near simultaneously, that provide a useful land-mark for this essay. One was B.F. Skinner's **Beyond Freedom and Dignity** (Skinner 1971); the other was Jacques Monod's **Chance and Necessity** (Monod 1971).

P. Grobstein (Deceased) (✉)
Center for Science in Society, Bryn Mawr College, Bryn Mawr, PA, USA

© Springer International Publishing AG, part of Springer Nature 2019
J. A. S. Kelso (ed.), *Learning To Live Together: Promoting Social Harmony*,
https://doi.org/10.1007/978-3-319-90659-1_14

Skinner drew on a life time of research in behavioral psychology to argue that human behavior was nothing more and nothing less than the product of the individual experiences that individual human beings had during their lives, and that with deliberate and careful control of those experiences it would be possible to achieve ideal human societies. Monod, on the other hand, wrote as a pioneering molecular biologist. While he himself made no sweeping claims comparable to those of Skinner, he was read by many as implying a different but comparable challenge, that human beings were nothing more and nothing less than the product of their genes. Monod was attacked as a proxy by those (including Skinner supporters) who felt that it was trivializing humanity to attribute its characteristics to genes, and Skinner was attacked by those (including Monod supporters) who similarly felt there must be something more to humanness than individual experiences. The deeper question posed by the two books went more or less unnoticed at the time. Suppose genes and experiences both affect behavior? Is that all there is to humanness, or is there something more (Grobstein 1991)? And if so, what? Must one fall back on spiritual or religious traditions to give us something more than our genes and our experiences, or is there something else that can be explored by empirical science?

If the brain is indeed the underpinning of humanness, as seems increasingly to be the case (Crick 1994; Grobstein 2002; Ramachandran 2003), then these questions ought to be approachable through empirical studies of the brain, and that has indeed turned out to be the case. That genes affect the brain was the subject of classical work (cf. Sperry 1956). And more recent work has made it abundantly clear that experiences do so as well (Kandel 2006). The human brain is influenced both by genes and by individual experiences, including those, like language and values, acquired by interpersonal and cultural influences. Given findings on the brain (and in other realms), there is no longer a nature/nurture controversy. The brain is proving to be the nexus of all influences on behavior.

In terms of the broader concerns of this essay, two further implications of this work are worth making explicit. One is that no two brains are identical. Given the interplay of two different sources of variability, genetic and environmental, and its own enormous complexity, the brain of every living human being is unique. Indeed, there is no possibility that the distinctive brain of any individual has existed in the past or will exist again in the future. The other is that every brain is itself continually changing. The brain is not fixed at birth, or at any later point in life, but is instead continually being altered in one way or another by its ongoing activity.

The brain gives each of us a unique and continually revisable self. But the question still remains of whether that self is simply a product of genes and experiences or whether there is something more, some additional features of humanness that can be identified from empirical studies of the brain?

The Harvard Law of Animal Behavior: Intrinsic Variability

"Under carefully controlled experimental circumstances, an animal behaves as it damn well pleases" is an ironic complaint of graduate students (and senior investigators) studying animal behavior (Grobstein 1994). And it is one that echoes day to day experience, outside the empirical laboratory, of anyone trying to predict the behavior of another person, or of themselves. We all have a sense that, irrespective of how well we know people (including ourselves) we can never fully predict behavior (including our own); there is always an element of surprise.

Could there in fact be something beyond genes and individual experiences with the outside world that influences behavior, in ourselves and other organisms? Recent empirical research on the nervous system suggests that indeed the answer is yes. Successive identical inputs to an isolated nervous system yield responses that differ each time (Grobstein 1994), with no evidence that successive responses depend in any way on previous ones. This sort of observation is common in studies or the nervous system of all organisms, but has tended to be ignored or dismissed on the grounds that the observations are not sufficiently "carefully controlled". increasingly, though, neuroscientists (Grobstein 1994; Carpenter 1999; Glimcher 2005; Maye et al. 2007) are beginning to take seriously the likelihood that there exist, within the nervous system itself, processes of "intrinsic variability", i.e. mechanisms that introduce a degree of randomness in nervous system function, so as to assure that "under carefully controlled circumstances" (i.e. identical genetic and experiential factors), the responses of organisms (including humans) to a given input will exhibit some random variation from occasion to occasion.

"Intrinsic variability" may seem to many like a cause without an explanation, perhaps verging on an appeal to the spiritual, but it is not in fact so. As the physicist Erwin Schrodinger pointed out many years ago...

> we know all atoms to perform all the time a completely disorderly heat motion, which, so to speak, opposes itself to their orderly behavior and does not allow the events that happen between a small number of atoms to enroll themselves according to any recognizable laws. Only in the co-operation of an enormously large number of atoms do statistical laws begin to operate... All the physical and chemical laws that are known to play an important part in the life of organisms are of this statistical kind; any other kind of lawfulness and orderliness that one might think of is being perpetually disturbed and made inoperative by the unceasing heat motion of the atoms (Schrodinger 1943).

To put it differently, some element of randomness underlies all regularities observed at the level of biological activity. That an element of intrinsic unpredictability should be present in brain function is not surprising. Indeed it would be surprising if it were not the case.

From this perspective, the issue is not whether intrinsic variability is real but rather why it exists, i.e. what adaptive biological function it might serve. Contemporary thinking suggest a number of possibilities, including the need of organisms to be unpredictable to one another. In the present context, however, what is of most interest is that some capability to generate outputs not fully determined by inputs provides a mechanism to acquire new information, to learn. Nervous systems seem to have an inherent capability to "try out things", to explore by generating unexpected and novel outputs to "see what will happen". And to learn by comparing what happened to internally generated predictions/expectations (Galistel 1980; Kelso 1982).

If nervous systems in general have this inherent capability, so too do humans. We are, by virtue of our exploratory capacities, capable of things that go beyond those made possible solely by our genes and our experiences. This creative learning capability, involving action and comparison of resultant with expectation, is as much a part of our humanness as genes, experiences, and, for that matter, neurons and synaptic interactions among them.

Distributed Processing: Society of Mind

we can make versatile AI machines only by using several different kinds of representations in the same system! This is because no single method works well for all problems; each is good for certain tasks but not for others. (Minsky 1986)

It is not uncommon for human empirical research Lo discover what has previously been revealed as important adaptive principles by the undirected processes of evolutionary change. It is increasingly clear that the human brain, like nervous system in general, has a "modular architecture", (Fodor 1983), i.e. that it consists not of a single master processor but rather of a set of more or less independent parallel processors, each of which has representations related to particular tasks and each of which has the capacity to update its representations with regard to those tasks (Galistel 1980; Kelso 1982). "Multiple intelligences", may be a relatively new idea with regard to psychology and education (Gardner 1983) but it is, in evolutionary terms, an old idea. Our internal sense of ourselves as single, coherent entities notwithstanding, we each consist of a large array of relatively special purpose processors, each of which is, in particular contexts, able to provide an informed and adaptive response to particular circumstances, often more rapidly and better than that which our coherent and experienced selves can provide (Gladwell 2005).

An ability to see and recognize others and ourselves as a "society of mind", rather than as single coherent individuals, is clearly important in acknowledging the different and distinctive capabilities of individuals in educational and other contexts. But it has significant wider implications as well. Among these is the importance of developing an ability to not only accept but actively value the idea that we are not always internally coherent (Rorty 1993), that internal incoherence in fact provides both the motivation and the grist to create new ways of making sense of both ourselves and the world around us (Grobstein 2004, 2005a, b). This valuable incoherence too is a part of the new vision of humanness that empirical research on the brain offers.

Part IV
Education and Culture

A Contrarian View of How to Develop Creativity in Science and Engineering

Martin L. Perl

I look forward to a deeper and more quantitative understanding of creativity as the science of brain research becomes more steadily and securely based in physiology and biology. I am an engineer turned physics experimenter and, at present, my crude and qualitative thoughts about creativity arc based on my experience in teaching, research, and working with colleagues in Silicon Valley and academia.

My interest is in creativity in the competent engineer and scientist, not in the star performers. I actually don't believe there is much to learn from analysis of their qualities (Simonton 2004) and I am tired of reading that Kekulé discovered the ring structure of benzene through a dream. Many colleagues and friends have told me about nightmares they have had about their research but not one has mentioned a productive technical dream.

I haven't found statistical and demographic studies of productivity in science (Stephan and Levin 1992) useful for understanding or teaching creativity. Austin's *Chase, Chance, & Creativity,* (1978) has been more useful to me than most, but I don't agree with his emphasis on chance.

Creativity is a broad aspect of humanity, but in engineering and science has constraints that do not exist in other creative activities such as music and art. An improvement of computer architecture, a discovery of a new medicine, a new understanding of the behavior of black holes, an improvement in gasoline engine efficiency -all are creative feats that are nonetheless limited by the laws of nature. We may be wrong about a law of nature. For example the consensus that energy producing, cold fusion is not possible is based upon our present understanding of thermodynamics and nuclear physics. The present understanding may be wrong, but that has to be demonstrated by consistent, quantitative experimentation.

M. L. Perl (✉)
Stanford University, Palo Alto, CA, USA

There is increasing pressure on technical education at the undergraduate, graduate and young professional levels as the amount of knowledge in engineering and science steadily increases. There is even pressure on the old professionals (Perl 2007). The usual response is to move college level courses into high school, move graduate level courses into college curriculum and extend the sophistication and specialization of graduate courses. This is harmful to technical creativity and in this paper I put forth a contrary educational approach.

Proposals

I have two proposals:

(a) Reduce the number of courses required for undergraduate and graduate degrees in engineering and science.
(b) Change the nature of laboratory courses and Ph.D. research so that the student has the freedom to try out her or his own ideas, with the expectation that they will make mistakes and will both expand their creativity and learn more by doing.

Basic Abilities and Skills for Technical Creativity

There are some basic abilities and skills that you must have for creativity in engineering and science.

Competency in Mathematics

You must be competent in mathematics even if you are in a field where mathematics is secondary. But you don't have to be a mathematical genius. The mathematical level should be that of a book such as Kreyszig (1988). You don't have to carry the properties of Bessel functions in your head, but you should recognize when your calculations need Bessel functions and know where to find their properties.

Imagination

Imagination is crucial to creativity in engineering and science, imagination within the constraints of known physical laws, experimentation, feasibility and practicality.

Begin with the far reaches of imagination at the science fiction level. Then gradually apply the constraints of physical reality. Beveridge in his *The Art of Scientific Investigation* has a marvelous chapter on imagination (1953).

Visualization

In engineering and scientific work it is crucial to be able to visualize how the work can be accomplished (Ferguson 1993). The intended work might be the invention of a mechanical or electronic device, the synthesis of a complicated molecule, the design of an experiment to evaluate the efficacy of a new drug, or the modeling of how proteins fold and unfold. There are many ways to visualize the development of your idea. I draw pictures and do rough calculations in my notebooks. Some primarily use a computer. Others make models. Still, others just carry out the visualization in their heads until most of the details have been worked out. If you are working with others, intermediate technical notes and meetings are necessary. If I am thinking in the wrong direction I prefer to know sooner rather than later.

Hands-on and Laboratory Skills

When choosing what you work on in engineering and science, honestly evaluate the extent of your hands-on and laboratory skills. Are you good with tools, with repairing equipment, or perhaps with using a microscope? You cannot be creative if your daily work involves activities that weaken your confidence and self-esteem. You can still do design work or theoretical work. Or, if you want to participate in the hands-on world, find a partner or a group with which to work.

My Ph.D. thesis advisor, Isidore Rabi, was given the Nobel Prize for his experimental work, but he had few hands-on skills. His graduate students were afraid to let him get close to their apparatus. When he came to the laboratory we immediately engaged him in conversation at the door, hoping he would get bored and leave. In spite of his hands-on limitations, he had a deep, mysterious way of understanding and visualizing experimental work.

Computers

Computers have changed the world of the engineer and scientist. Learn to use a packaged general computing program such as Maple, Mathematica or MATLAB. I use Maple because my friend Marvin Weinstein is a Maple expert and I can always go to him with problems. I find the best way to make progress in computing is to have an expert as a resource. The Internet is a great time saver for looking up references and reading papers. It is also great for looking up facts such as the properties of Bessel functions, but don't try to learn Bessel functions from the Internet -for that you need a textbook. The curse of the Internet is email. I look at mine no more than once a day and keep my finger on the delete key.

Developing Good Ideas in Engineering and Science: What You Should Do

Good Ideas

Good ideas in engineering and science take many forms including: simplifying a consumer electronic device, improving a surgical procedure, discovering something new in topology, or developing a technology for finding life on planets outside the solar system -and the list goes on and on. Bringing a good idea to fruition brings pleasure and recognition to the practitioner, as well as career advancement and money. And so it is fulfilling on many levels to get a good idea and make it work.

But for every good idea, expect to have five or ten bad, wrong or useless ideas. This is my experience from fifty years of observing the creative work of the engineers and scientists that I know. Sometimes the bad idea does not survive a conversation or some clear thinking over a weekend. But sometimes you get to the stage of building a prototype or an experiment or publishing a paper before you realize it is a bad idea. And sometimes the thing is already built.

Gilbert's *The World's Worst Aircraft* is full of horrifying example of bad engineering ideas (1975). In science sometimes it can take a century for a bad idea to be defeated; phlogiston and the electromagnetic ether are examples. There are many reasons why ideas are bad -perhaps it violates physical laws, or a competitor has a better product based on the same idea. The only way to proceed in creativity is to use 'patience and fortitude' (LaGuardia 1930) in looking for the good idea.

Unfortunately most histories and biographies in engineering and science neglect the abundance of bad ideas. This is partly due to hero worship and partly due to the writer not being an engineer or scientist. They just don't know about all of the bad ideas, and nobody bothers to mention their abundance. I hope I don't make too many enemies by pointing out that books about Einstein's work usually err in this direction. After enthusiastically discussing his stupendous early work they spend little space on his many erroneous ideas on unified field theory after 1925, and the fact that he ignored important strong and weak forces. For example, Isaacson in his recent, popular 500 page book on Einstein (Isaacson 2007) devotes only a few pages to ideas that didn't work. Where is the young engineer or scientist to learn about the prevalence of wrong ideas in the work of great engineers and scientists?

Edison's laboratory style is a marvelous example of the success that can come with acknowledging that most ideas turn out to be useless and yet continuing to give everything a try. But my favorites, by far, are the entertaining, overblown accounts or Francis Jehl (1936).

Finally, don't try to hide a wrong idea or wrong results, for as Medawar (1979) says, "The important thing is not to try to lay down some voluminous smoke screen to conceal a blunder".

Good Ideas and the Technology You Use

To get a good idea you must be immersed in some technology: biological, electrical, mechanical, or mathematical. You must be interested in, and perhaps even enchanted by some of the technology, software, or mathematics you use. Then the bad days are not so bad. Another advantage of being enchanted by a certain technology is that you will be more likely to think of improvements and variations. You should be fond of the technology that you use, but not so much in love that you are blind to the possibility that there may be a better way. Also, avoid the natural tendency to ignore technology that is 'not invented here'.

Colleagues

Colleagues who are supportive and helpful will aid in the development of a good idea and shorten the time you spend on bad ideas. And as emphasized by Medawar, it is important to (Medawar 1979) remember that technicians are colleagues, too.

Getting Good Ideas in Engineering and Science: What the Educator Should and Should Not Do

We Overeducate

As engineering and science keep changing and expanding, we educators keep pushing the students to learn more and more. Repeating my introduction, we move college level courses into high school, move graduate level courses into the college curriculum and extend the sophistication and specialization of graduate courses. The student's time is filled with studies, homework, and testing. There is little time for the student to play with ideas, to dream about discoveries and inventions. We overfill the student's time and the student's head. Below I will discuss two proposals that will alleviate this.

Reduce Requirements for Degrees

I propose that the course requirement for undergraduate and graduate degrees be reduced to basic subjects. For example, the course requirements for a physics doctorate should be limited to advanced courses in classical mechanics, quantum mechanics, electromagnetism, quantum field theory and statistical mechanics, as well as intermediate level courses in solid state physics and elementary particle physics. The students might take other courses such as cosmology, string theory, advanced fluid mechanics, or biophysics depending on how they want to spend their time.

Teach students to learn as they go in their work or in new projects. Teach them that they don't have to do extensive study to move into new technical areas, they can learn a subject or a technology as needed. Emphasize learning by doing.

Change the Nature of Laboratory Courses and Ph.D. Research

I propose that laboratory courses be revised so that there is an emphasis on process and problem solving rather than finishing prescribed experiments. Allow the students to try their own ideas and to make mistakes. As an undergraduate in chemical engineering I hated chemical quantitative analysis, and the finicky methods I learned were soon made obsolete by the march of technology. It would have been much better if I could have set my own analysis problems.

There is an overemphasis on 'original research' as a requirement for a Ph.D. The work is usually part of a larger, ongoing research program. It is primarily training in R&D. Reduce the pressure on the doctoral student by not depending on them to justify the professor's salary or grant. Encourage students to try out their own ideas, make, and learn from their own mistakes.

Acknowledgements 1 am very grateful to Professor Veljko Milutinovic of the University of Belgrade and Professor Hironori Fujii of Tokyo Metropolitan Institute of Technology for stimulating discussions on the subjects of creativity and innovation.

This work was partially supported by Contract DE-AC02-76SF00515 with the U.S. Department of Energy.

References

Austin, J. H. (1978) *Chase chance. & creativity.* New York: Columbia University Press.
Beveridge, W. I. B. (1953). *The art of scientific investigation.* London: William Heinemann.
Ferguson, E. S. (1993). *Engineering and the mind's eye.* Cambridge: The MIT Press.
Gilbert, J. (1975). *The world's worst aircraft.* New York: St. Martin's Press.
Isaacson, W. (2007). *Einstein.* New York: Simon & Schuster.

Jchl, F. (1936–1941) *Menlo Park Reminiscences*. Dearborn: Edison Institute. This three volume edition is out of print but can be obtained from used book dealers. Dover Publications, New York, has republished the first volume.

Kreyszig, E. (1988). *Advanced engineering mathematics*. New York: Wiley. This is a twenty year old book but this demonstrates that the mathematics we usually need changes slowly.

Medawar, P. B. (1979). *Advice to a young scientist*. New York: Harper & Row.

Perl, M. L. (2007) On learning in engineering and science when old, SLAC -PUB 12701 (2007), to be published.

Quotation from Fiorello LaGuardia, Mayor New York City during the 1930s Depression years.

Simonton, D. K. (2004). *Creativity in science*. Cambridge: Cambridge University Press.

Stephan, P., & Levin, S. (1992). *Striking the mother lode in science*. New York: Oxford University Press.

Martin L. Perl was professor emeritus at the Stanford Linear Accelerator Center, Stanford University. He was awarded the Nobel Prize in Physics in 1995 for the discovery of a subatomic particle, the tau lepton. Sadly, Professor Perl passed away on September 30, 2014.

The Ever More Pressing Problem of Science Literacy

Leon Lederman

The causes of science illiteracy—the superficiality, misinformation, ignorance, and downright hostility toward science we encounter at virtually all levels of school and throughout adulthood—are numerous enough so that rooting out the problem will require a complex and sustained effort. But this is a war well worth fighting, and it is winnable as long as we choose our battles well, employ intelligent strategies and hang in there for the duration. As Winston Churchill exhorted his compatriots: "We will fight on the beaches, we will fight in the cities, we will never surrender!" (from "Getting High School Science in Order" by Leon Lederman.)

Before I talk about education I want to talk about the state of science today. A new study on the status of R&D reflects gloom; projections of support for basic research are very dismal. Research in this country is done in three places: industry, universities and national laboratories. Industry is getting out of research—the once-great industrial laboratories are no longer doing basic research or even basic applied research, and Bell Laboratories is down to about 30% of what it once was. Many of these laboratories were incredibly productive, but there's a very strong move for industry to abandon basic research altogether. Federal support of basic research has not been very profound. If you look at non-defense research, you find the U.S. is way, way down the list compared to many other countries. This has been alarming even to editorial writers; a New York Times editorial talks about crippling American science, and the Washington Post headlines an article with "Squeeze on Science." Recently CEOs of the sixteen major U.S. high-technology corporations took a full-page ad in the Washington Post pleading with Congress to support basic research. Unfortunately, they only took one ad.

Superimposed on this depressed situation are awesome possibilities. There are emerging technologies which could add a trillion dollars to the gross national product over the next decade: superconducting materials, optoelectronics, artificial intelligence, high-tech ceramics, the whole range of molecular biotechnologies and so on. There are many of these emerging technologies, and it's not at all clear what

L. Lederman (✉)
Fermilab, Batavia, IL, USA

© Springer International Publishing AG, part of Springer Nature 2019
J. A. S. Kelso (ed.), *Learning To Live Together: Promoting Social Harmony*,
https://doi.org/10.1007/978-3-319-90659-1_16

fraction will be captured by U.S. industry. The Japanese have taken an interesting approach. An official report of the Japanese government reviewed the Japanese economy with disappointing projections, and suggested that the way to reverse the situation would be to increase support of basic research. As a result official Japanese policy is to double the country's research budget over the next five years. Way back in 1922, one of the world's great futurists, H.G. Wells, wrote that "Human history becomes more and more a race between education and catastrophe." That is certainly true today. I want to talk about two particular activities in education, but let me first set the scene.

A National Initiative in Education?

In the rhetoric about education and education reform, especially among state governors and presidents, you hear over and over again words like "we need a change and it must be 'bold,' 'radical,' and 'break the mold.'" I want to raise an issue that might be controversial, certainly should be controversial. Our founding fathers in their almost infinite wisdom gave responsibility for education to states and localities and, therefore, the federal role is relatively minor. It includes supporting education research, and importantly, providing a bully pulpit. But how could the founding fathers have anticipated the importance to the nation of a science-literate workforce, and a science-literate citizenry? Now we don't depend on South Dakota to defend the United States; we give that responsibility to the federal government. Should we not at least examine the proposal that a national initiative in education is called for? This came to me after attending several so-called "renaissance weekends," interesting gatherings of CEOs of major corporations and government officials and journalists—wise people. After listening to the various problems facing this country, I made a list. Every one of these problems, major problems, ultimately has education as a vital component of its solution. You could make your own list: decay of the cities, crime, drugs and the increasing gap between the rich and the poor in the United States. (We're catching up with Brazil which has a ratio of 100:1, the average income of rich versus poor, the top versus the lowest levels. In the Pacific Rim, incidentally, the ratio is about 5:1.)

Other problems include those of the environment; the problems of energy sources which are benign for the environment; endemic diseases like AIDS; the whole problem of population, a major problem which was important enough to collect the leaders of some 130 nations in Cairo in 1994; industrial accidents which include oil spills; air safety; dwindling research budgets, which I already mentioned; racism and gender bias; junk science; the worldwide growth of fundamentalism, from the Taliban rulers of Afghanistan to the hardliners in Israel and the Arab world to creationism in the United States and so on. Despite the importance of education in solving these problems, the U.S. is not doing very well, this from international studies of science and math achievement. One global report called the problem one of focus. Let me address the issue in a general way. Visualize a circle: when I get into educational debates, I listen to many strong opinions about the origin of problems which begin with preschoolers, persist throughout life, and revisit the next generation. Some say the problem starts in preschool because that's where the attitude of children towards school is shaped; or the problem is in K–8, the problem is in high school, the problem is in the colleges—or the problem is that the general public is glued to the television set. I decided to finesse this debate by drawing a circle and saying, as in an elementary electricity course, that if you don't have a complete circuit, no current will flow. You need to fix all of these problems, because the graduates of our high schools and colleges come the full circle and become voters and parents. You might have access to these voters and parents for a general program on "public understanding of science," and you can use TV and radio and op-eds and museums and malls, and you might even put science stories on cereal boxes. Then there are movies which portray scientists either as nerds ("Honey, I shrunk the kids") or monsters ("Tomorrow we destroy the world"). That's Hollywood.

Reforming Chicago Schools

So you have all these problems, and what I want to do is spend a little time on the good news, which is that they are not unsolvable. Back in 1989 I had just moved into the city of Chicago from the comfortable suburbs to become a professor at the University of Chicago. The city passed a dramatic school reform bill which decentralized Chicago schools, giving us 540 little corporations, each with a CEO and elected board of directors. The CEO is the principal, still called "the principal," but the board of directors has control. It can fire the principal if he or she doesn't "make money," that is, improve student and educational performance. In the euphoria of all this, listening to colleagues who had been involved in passing this school reform bill, it was very clear that the school system needed some friendly intervention. One of the major problems was the poor preparation of primary school teachers, K–8, but mostly K–6, for teaching math and science. There was a perception that these primary school teachers were horribly underprepared, and it turned out to be true: it's not possible to exaggerate their lack of preparation. It's not the fault of the teachers, because we have found that they will work enthusiastically to become better teachers if you give them opportunities. In order to attract candidates to teach primary school, teachers' colleges practically posted signs saying we'll protect you from math and science, just come.

And so they were protected until they were classroom teachers, and suddenly at two o'clock in the afternoon, full of their own insecurities, they had to teach math or science. We thought we could help these teachers, so we created something called the Teachers' Academy for Math and Science, a freestanding institution for professional enhancement. We got support from the Department of Energy, a curious place to get support, but we did. We have 24,000 teachers for 410,000 students, and—to recite a litany typical of the cities—68% of the students come from families under the poverty level, and they scored in the lowest one percent on any of the national tests. Chicago has crimes and gangs and drugs and old buildings and multi-ethnic backgrounds and poor parental support, you name it. Nevertheless, we thought that if we could help the teachers, they would be at least one component in a solution.

After Extensive Teacher Training, Progress

We developed an in-service program with a staff, and we use all the new pedagogy and a hands-on, inquiry-based, process-rich curriculum. We steal programs from everywhere, without any bias. We take teachers out of classes, replace them with substitutes, and give them a lot of science; we also work with them on weekends. We found out that all of this pedagogy must be distributed over a minimum of three years, more likely four, and costs about $10,000 per year per teacher. As I said, we teach them during class time, summers, weekends, evenings. We do this systemically and get the parents and local community groups involved. Assessment and evaluation is a new profession, but we've used as much as anyone could ask. One measure is the State of Illinois math exam, a statewide exam that's been given for many years. It has a maximum score of 500, and the state average is about 270. City schools began to catch up in a dramatic way after two years of work with us. There's a long way to go: we started with the worst, the poorest schools in Chicago; we deliberately took the schools with the highest poverty levels and the worst reputations for academic achievement.

The results show an improvement after the schools have been in the program for at least two years, but it's even more dramatic after three years. My dilemma is that we really need another four or five years to see if all the things that we're not doing will overwhelm the things we are paying attention to. Hands-on science is very engaging. I've seen teachers get the children to calm down when they say, "Kids, if you don't settle down, we won't do science." And they sit up, and it's quite a thing to hear these youngsters with their street English say, "Hey man, what's your independent variable?"

I want to discuss the high schools next, because we all share a professional interest in secondary education. You know that a consensus is emerging on national standards for math and science. The math standards have been around for a while; the first national standards for science were the AAAS benchmarks, and these were reiterated by standards written by the National Research Council of the National Academy of Sciences. So we are getting a good consensus of what kids should know in second grade, in fourth grade, in sixth grade and so on. Another trend is the increasing number of school systems that are beginning to insist on three years minimum of science and math.

The City of Chicago adopted this standard in principle a couple of years ago. New York City adopted the standard, and the state of Pennsylvania requires three years. I think the trend is very encouraging, and we're heading towards a system in which all high school students take three years of math and three years of science as part of a core curriculum. This is not for scientists, this is for citizens—all high school graduates. The Chicago schools realize that they don't have the teachers or the laboratories to meet this standard, and they hope to phase it in by the year 2000.

Three Years of Science—In a Coherent Order

My main point: if we require three years of science and math, and if nationally accepted standards of math and science exist, then why not seize the day and create a coherent science curriculum? Let's call it Science I, II and III to try and smooth over the disciplinary categories. And let's try to make it coherent, because you now know that there are going to be three years of science. All too often in today's schools—some huge fraction of the schools—students begin with biology. Since they have had no chemistry or physics, the biology has to be qualitative with lots of memorization, stultifying for a subject as dynamic and exciting as modern biology. I will fully cede the second half of the 20th century to biology; the first half was physics, of course, and chemistry is there all the time.

When today's kids finish biology, they put a period at the end of the exam and it's forgotten. Some of them will go on to take chemistry, and then certain areas of biology might begin to make sense, but because they had no chemistry when they took biology, the knowledge gained is less than it should be. About 20% go on to take physics, mostly as seniors (the ones that are going to shrink the kids or destroy the world).

There's another way of doing it that makes sense and which corresponds to the way science works. Science works in a hierarchy. It's a pyramid with mathematics at the base. Physics requires mathematics and is second (mathematicians, on the other hand, don't care whether physics exists or not; their discipline isn't dependent on physics). So physics sits on top of mathematics as a necessity. Chemistry sits on physics because chemistry involves the interaction of two atoms to make a molecule, and atoms are the bag of physicists. Chemistry depends on physics. Modern high school biology—not the qualitative biology which is very beautiful and should be emphasized in K–8—depends on molecules like DNA. How can you study biology and not mention DNA? And how can you understand DNA if you don't know what a molecule is? And how do you know what a molecule is if you don't know what an atom is?

So the pyramid of science is suggestive of the way we should rearrange our high school curriculum to start with ninth grade physics. Ninth graders still don't know much algebra. If we do impose this pyramid scheme, it will put pressure on the primary schools to teach some algebra in eighth grade; Science I will then be mostly conceptual, mostly physics, with excursions to chemistry where it is relevant. But these classes will hand over to Science II—which will be mostly chemistry—kids who know what an atom is, and who will understand how an atom is held together.

They will also understand the process by which two atoms decide to lock onto—or avoid—each other, which is called chemistry. Then in the study of molecules, which kids will become familiar with from chemistry, they will be prepared for molecular biology—gene structure, proteins, DNA and so on. The pedagogy is that they can begin to use physics in the study of chemistry, and both physics and chemistry in biology. And with this, you have coherence, you can talk to one another. We must have physics teachers talk to the chemistry and biology teachers, and even know a little bit about those other subjects. Biology teachers might say, "O.K., today we'll learn about photosynthesis. Let's brings that physics teacher into remind us what a photon is. And then our chemist will tell us about photo-induced chemical reactions, and then we'll talk about photosynthesis." You begin to stress the unity of the sciences, the conservation laws which are valid in all subjects, and symmetries, vibration theory and so on. You really have a three-year subject which not only covers the science in a pedagogically sensible way, but leaves you time to talk about science and society. This could include material on the qualities of science; the process of science; the delicate balance between skepticism and respect —important because these are citizens. We are training people to be voters and citizens, and to be intelligent decision makers. We want them to think scientifically.

Promoting Education Reform

As activists we formed an organization which we call ARISE, American Renaissance In Science Education. We had a meeting last September, and now have a steering group with officials from the National Academy, the AAAS and the National Science Teachers Association. We're beginning to think about how to change a system—biology, chemistry and physics—which was established in 1893 by a committee of ten, chaired by a Harvard president. Maybe that committee's logic was that the course order should be alphabetical. Seriously, I don't know what the logic was, but 100 years ago it probably didn't matter. It certainly matters now, and this change has been resisted for 100 years. We have to respect the difficulty in getting this done in all the high schools in the United States; we even found resistance among the physics teachers visiting Fermilab last summer because they didn't want to teach ninth graders. In the fourth year of high school students can, if they are college bound to do science or engineering, take advanced placement courses in any of these subjects, or even begin an advanced placement in the third or second year if the high school is qualified to offer it. One can go to a liberal arts college with enough science so that the college can't get away with "rocks for jocks" or whatever passes for the minimum science requirement at too many institutions. At some schools you can get a bachelor's degree without changing any of the misconceptions you had regarding science as an entering freshman. Or you can go to work. And if you go to work, you might want to take courses in technology, computer science and all the things that will help you. With a base like this there are possibilities in earth science, environmental science, all kinds of interdisciplinary subjects which could compose an important curriculum. That's the ARISE program.

There are going to be obstacles. I've made a list of the obstacles, because you have to understand how hard it is to make a revolutionary change. A 100-year record of resistance to change, that's obvious; it requires a strong, enthusiastic, visionary, youthful, well-financed program. It's going to be expensive: you need computer terminals, probably new labs and new books. You need continuous professional enhancement, you need time for teachers to talk to each other. And we've got to get the parents involved, because their biggest objection will be, "If you make this change my kid won't get into Yale," and you'll say, "Well, that's lucky."

One of the first things I did was to get an advisory group to verify that all this was not a plot by a physicist to torture ninth graders by making them study physics. I got 25 Nobel laureates in biology, chemistry, physics, all the science subjects in which they give Nobels, to certify that this makes sense. I'm now working on the customers of the high schools—prospective employers—the CEOs of major corporations, and maybe a few military people.

Let me address the subject of the universities and their role in this reform. Perhaps the key to fixing the educational system lies in the structure of the universities, blending the disciplines. I have lot of respect for physics as a discipline, and also for chemistry and biology, but I don't like walls. Instead of walls there should be shrubs and little paths between disciplines so you can go comfortably back and forth. Unfortunately, that's a very unrealistic vision of the university which is crushed by conflicts between the soaring imagination of gifted scholars and the rigid, unyielding turf established by departmental politics.

Community Support Is Vital

We can't meet the goals of fixing the sciences in the high schools or the elementary schools without making lots of other changes. We can't affect priorities unless we can get teachers, parents and school administrators on our side, and ultimately we need the general public as well. And I think we can't fix the natural sciences without addressing the humanities and the social sciences, a recognition of unity of knowledge. The two cultures gap is only getting worse, yet in this process of changing the high schools we have to incorporate the social sciences and the humanities. When you study history you study the kings of England, and you study Napoleon and Genghis Khan, yet you never study Faraday. But Faraday changed the lives of people on the planet more than Genghis Khan or all the kings of England rolled into one. He had a bigger influence, but this fact is somehow not taught in history class.

If you agree with me that education is key to solving the social problems we listed earlier, then we have to raise education's priority in this country to an entirely new level. After WWII, science was great; it had learned how to make deafeningly loud noises, and it did all kinds of things in the war. There was a wave of euphoria which elevated the status of the educational system for a while, then faded. Then came Sputnik which again established new priorities for education. The textbooks, physics, chemistry and biology, improved, but five years later those books faded in their influence; you couldn't get them, and the new ones were watered down. The new books were not produced with sensitivity, and didn't explain what they were trying to do. Also, they focused on the pipeline problem rather than on raising the level of the general student.

I'd like to see our priorities for education roughly equivalent to those we might establish in our small wars. We know that there are some good schools. Why can't all of our schools be as good or better than our good schools? It will take an unfettered imagination. This country occasionally establishes national commissions, and in fact the one impaneled by Terrance Bell created the 1983 report, "A Nation at Risk." With purple prose and military metaphors, it sounded the alarm over the quality of U.S. education. It said we had committed unilateral "education disarmament," and if any country did to us what we did to ourselves, we would have nuked them. Maybe exaggeration was needed.

Drawing Plans for Action

Government is probably not the place to organize a national commission like that one. It should be organized by an organization like Research Corporation, for example, which could impanel a group of 20 or so wise people: educators, scientists, CEOs of major corporations and maybe an occasional university president. The commission should take a year or two to examine the obstacles to achieving a superb education. Why can't the U.S. be number one? What are the obstacles? Are they the teachers' unions, the parents, the universities which refuse to raise their standards, refuse to hold the high schools to higher standards? Let's find out what attitudes, laws, finances and institutions need to be changed. This commission could issue a report as dramatic and more far-reaching than "A Nation at Risk." It would be a call to action, and would tell us what to do to make a dramatic change in our educational structure in order to be number one by the year 2050. The action plan should be handed to the president of the United States. I'd say, "Mr. President, here's a plan which will give us superb education; this is what it would cost, this is who we have to convince, and this is how we should go about doing it." It's an enormous challenge, but after all, somebody invented free public education. Somebody thought of land-grant colleges, and they have had a tremendous impact on the development of the United States. Someone invented the superb U.S. graduate schools which are still the best in the world. If we give primary and secondary education this kind of priority, I think we can succeed.

I want to conclude with a story that my mentor, I.I. Rabi, always told. Rabi's little essay was called "Plato's Academy." Plato had an academy for preparing young men for entry into the world, a world perhaps as troubled as ours seems to be. Rabi thought it would be interesting to put a fresh, bright graduate with a B.A. from an Ivy League institution in a time machine to attend Plato's academy. What would our graduate be able to tell his opposite number at the dawn of recorded history? Could this product of two millennia of human progress enlighten the Greeks on philosophy, the art of sculpture, painting or architecture? Could he demonstrate a superiority in poetry, drama, history or oratory? Would his language be more pure, his thoughts more elegant? Would he excel in athletic prowess, in grace of body? Would he be more prepared to fight for his country, to lead men into battle? One doubts whether our 20th century grad would make a brilliant showing in these fields. Indeed only a small fraction of our students could get by the door of the academy, because at the entrance was the inscription, "Let no man enter who does not know mathematics."

Leon Lederman (Nobel Prize in Physics, 1988) is Emeritus Director of Fermilab in Batavia, Illinois. At the Eighth Olympiad of the Mind, held at the US National Academy of Science in Washington, DC, he gave a passionate after dinner speech about his work on enhancing science and math education in the public schools of the United States of America. The style, not only the content of his speech is captured in this earlier presentation to high school science teachers and their mentors (adapted from "Science on the Frontiers", Research Corporation Foundation). Though one may not agree with his inimitably reductionist view of science, Lederman's message seems as prescient today as it did 20 years ago.

Brain Science: A Meta-Discipline for Education

Ron Hoy

Our conference calls attention to and acknowledges the remarkable discoveries about how the human brain works and their potential for improving social harmony in our troubled world. Of course, the process of education is a prominent theme throughout our sessions, and mine in particular. When we think about the role of brain processes in education and even more ambitiously, in social harmony, we are acknowledging that we are thinking about the process of thinking and learning-which take place within each of us individually, as well as how this learning affects our social behavior that comprises societal norms. It is in this sense that I refer to "meta-discipline" in my title.

The Processes of Teaching and Learning in the University

The Goal

The curriculum reforms that arc being implemented even as they are being contested that I'm most familiar with are in the biology and life sciences, and particularly with respect to biology's expanding interface with the physical sciences and mathematics. Consensus on educational goals is surprisingly hard to come by in the university community, but one goal, found in survey after survey, that is not contested by anyone is the ability to think critically. It may seem surprising that there's not a uniform definition of just what critical thinking is. So I'll give you two from authorities in their field. The former President of Harvard, Derek Bok, in his thoughtful book, *"Our Underachieving Colleges"* favors a definition from the American Philosophical Society:

R. Hoy (✉)

Department of Neurobiology and Behavior, Cornell University, Ithaca, NY, USA

e-mail: rrh3@cornell.edu

Purposeful, self-regulatory judgment which results in interpretation, analysis, evaluation, and inference as well as explanation of the evidential, conceptual and methodological considerations on which a judgment is based.

Another definition comes to us from the former president of the American Psychological Association and cognitive psychologist, Diane Halpern, who writes:

Critical thinking is the use of those cognitive skills or strategies that increase the probability of a desirable outcome. It is used to describe thinking that is purposeful, reasoned, and goal directed–the kind of thinking involved in solving problems, formulating inferences, calculating likelihoods, and making decisions, when the thinker is using skills that are thoughtful and effective for the particular context and type of thinking task... Critical thinking is more than merely thinking about your own thinking or making judgments and solving problems–it is using skills and strategies that will make "desirable outcomes" more likely.

Other scholars point out that critical thinking implies achieving an objective, or disinterested point of view, by applying a logic-guided deconstructive process to arrive at a conclusion that offers a dispassionate basis for decision-making or evaluation. The dispassionate part clearly means setting aside one's emotions to achieve clear-minded critical thinking. Really? We shall return to this point later.

The Practice

At university, whatever its immediate or long-term goals, the method of instruction has long been, and still remains the 50-min lecture; it has its pedagogical roots in antiquity. But there are reasons besides history and institutional inertia. The lecture is good for transmitting a large amount of information in a short span of time, and is personalized in the sense that the lecturer is, presumably, an authority who has critically culled off the very best thinking in the field, which would otherwise take months or years, working alone. The assumption is that the transmission of information occurs without loss—a process rather like downloading a set of files from the mind of an authority into the many minds of students, assembled in a room. Nowadays, especially in the sciences, the professor's information or data set is often presented as a series of Powerpoint slides, a practice which is also hotly contested and deservedly so. Teachers tend to love Powerpoint; students more often hate it. But as we learned last month, a good Powerpoint presentation on Inconvenient Truths helps win its presenter an Academy Award and even a Nobel Prize. This is one application of technology, for better or worse, in the classroom. But there is another and I think, much better application. Over the past decade, in the sciences at many colleges and universities, the effectiveness of the traditional lecture has been questioned, especially among those teachers who are most passionate about good teaching. They claim that reliance on lecture-based teaching does not promote deep learning in their students, especially when it comes to understanding concepts and solving problems. These teachers have developed a mix of practices called active-learning, in which students themselves necessarily become involved in their own instruction, during the class-hour itself. It can take the form of participation in peer groups, where 3–6 students form teams to work together to solve problems or develop reasonable hypotheses in tackling complex concepts. Active learning can also occur in large lecture halls filled with hundreds of students. This is enabled by wireless technology that permits the instructor to instantly "quiz" her students at any time, and display the results to the class—also instantly—in a manner that permits anonymity and is paperless. Clickers look and operate very much like your TV or DVD remote controllers. Each student purchases his or her own clicker for about $25 at the beginning of a term and brings them to every class throughout the semester.

How are they used? Instructors can pose multiple choice questions, answered A, B, C, D, at any time in the lecture or discussion and each student must answer by choosing (or clicking-in) her choice. The choices are transmitted individually to the instructor's laptop, for grading purposes, and the class's collective choices are displayed in the form of a bar graph on the big classroom screen, instantaneously. This lets everyone know where they stand—the teacher can go back and clarify a point that obviously confused the class, each student gets immediate feedback on her understanding of the material. It is important to note that no student is identified individually: the result or census that is projected depicts the decisions of the class as a whole as a frequency histogram. Class participation is 100% and no student needs to be called on in class to speak up-and fear embarrassment. The use of clickers is rapidly spreading and we may expect in just a few years, to have systematic assessments of their effectiveness, both from teacher and student viewpoints. This is a welcome change—*the use of clickers has not eliminated the teacher's lecture*—current usage is to use them to "check in" on a student's understanding of a lecture, moment-to-moment, or to record how well the students responded to a previous reading assignment meant to prepare students for the next lecture, or to simply know where the students stand in terms of preconceived ideas about material to be covered in that day's lecture. The success of clicker teaching is closely tied to thoughtful construction of clicker questions, because students tire and become contemptuous of simple-minded quizzing of factual knowledge. The aim of good clicker pedagogy is to help students think more critically by making them aware of their preconceptions and prejudices by giving them on-the-spot feedback.

There are many other practices that involve active learning that are used in today's classroom, from K-12 to colleges. This is beyond the scope of the present paper, but they can be found in any number of books, some of which 1 depict here. I want to return now to the question of what brain science is bringing to education in terms of our goals for better education and hopefully, global harmony.

What Are the Brain Sciences?

Broadly, it's what neurobiologists, cognitive neuroscientists, and psychologists do. Their concerns include how the brain generates acts of behavior, including cognitive acts, emotional acts, and movement acts, within an individual human. But humans are social, so how brains interact collectively, within societies, cultures, and importantly, in educational institutions, is also a concern. Importantly, there is a tacit assumption that brain equals mind. The collective content of that which we refer to as "what's on our mind" *emerges* from the totality of biological and physico-chemical activities that go on in our brains. Hence, what goes on in brains can be investigated and determined by the investigative procedures of science according to the laws of physics, chemistry, and biology. The trick is what's meant by emergence, but that's another entire issue. Nonetheless, brain science assumes that we can apply the reductionistic and deterministic methods of science to understand our brain, our mind.

Nature and Nurture

Understanding human behavior in scientific terms is not new to brain science. That goal gave rise to the discipline of Psychology. For about a half-century, psychology in the U.S., went about trying to explain behavior without recourse to what was going on in the brain at all. This was of course, Behaviorism, which considered the brain as *terra incognita,* and essentially unknowable.

In trying to explain human behavior, some psychologists had tried to parse the effects of biological endowment from those of the environment in which the individual grows up and matures. A goal was to come up with universal laws of human behavior but the nature-nurture confound persists in psychology, then and now. Brain science has its roots planted in biology, a discipline which has thrived on finding universal mechanisms that underlie life itself, common processes that unite bacteria to humans, especially at molecular, cellular, and genomic scales of life. Such a process suggests that brain science might succeed in parsing the effects of nature from nurture as they interact to shape human behavior during the life course, from infancy to adult, as has been so written in some quarters.

The Brain, Learning, and Education

You might think that one of the most active areas in neuroscience and cognitive neuroscience might be its application to education. But not so, education departments rarely interact with neuroscience departments, or even with cognitive psychologists in psychology departments. But that situation is changing, largely due to the efforts of some neuroscientists themselves who realize that in their studies of brain plasticity, recovery of function after stroke or injury, and how to slow the cognitive ravages of age-related dementia, that how humans acquire and retain knowledge and life experiences is an immediate and urgent priority at both ends of the human life cycle, childhood and old age.

Functionally speaking, learning can be divided into several sub-functions, all of which contribute to the end of storage in our memories. First, you don't learn anything if you aren't paying attention. Learning occurs in the conscious state and attentional resources are mobilized when salient sensory stimulation activates a neural system in the brainstem that is co-active with the conscious state. Stimulation of the senses is passed along other neural systems in the brain, including the thalamus, appropriate sensorimotor cortices, and the prefrontal cerebral cortex. These are the largest divisions of the brain and each can be subdivided by sensory modality and cognitive or motor functions. I'm trying not to get overly technical because the neuroanatomy of the brain is stunningly complex. It is not for nothing that one metaphor used to describe the brain is the internet and the mathematical analytic tools to study it, network theory and graph theory.

Functionally, the act of learning means getting and paying attention to the target, mapping the sequence of sensory inputs representing learning acts into brain space, and then storing it in the brain so that it can recalled at a later time. Since I mentioned a central goal of education as critical thinking (CT), we can unpack the definitions and try to reframe them in terms of brain tasks. Recall that implicit in the goal of CT is that it is dispassionate and objective—evaluating the facts or situation at hand to come to an objective decision. However, what's become clear from today's cognitive neuroscience is that, in reality, the emotions can never be factored out of our behavioral actions. The parts of the brain that add emotional coloration to our sensory perceptions and to our decisions, including and especially, the amygdala, is very much involved in the learning process. Consider the case of "math anxiety," which educators acknowledge and teachers strive to overcome in their students. Anxiety as a conscious state is mediated by the amygdala and other parts of the limbic system, that includes the hippocampus, which is critically involved in learning and making long term memories.

Whether we will ever be able to identify a "road map" or the neural connections that characterize the process of a mind engaged in thinking critically is an intriguing thought. But there are brain scientists who believe it possible that there may exist functional modules for cognitive processes that are represented by anatomically identifiable nuclei in the brain. There are hypotheses that claim there may be universal modules in the brain for language, mathematics, and even music. These ideas likely originated in neurology clinics where the effects of strokes, tumors, or accidents that can befall the brain and can cause striking kinds of cognitive deficits that present with remarkable specificity and predictability. Neuropsychologists refer to such catastrophic events as "experiments of nature." A perusal of this literature gives one the impression that such modular organization is a general organizing principle of functional anatomy within the human brain.

Powerful tools are now available for imaging the living brain. They are so non-invasive that they can be applied to purely experimental settings using paid volunteers as subjects. Functional MRI (fMRI) permits the investigator to observe the activity of a subject's brain while he or she is involved in solving puzzles, remembering events from their lives, or watching a video designed to draw out a subject's emotional reactions. fMRI is a more powerful technique than the post hoc method of functional inference after brain injury because it reveals the workings of the "normal" brain in almost real-time and with a good deal of anatomical accuracy, within the limits of the technology, which promises to get only better in the future. I'd like to present the findings of some fMRI studies that bear on subjects of interest for this symposium.

Recently some relevant studies have appeared in the *Proceedings of the National Academy of Science* that raise important brain and behavior issues. Consider the totality of human intellectual activity. When we think about critical, logical, and precise modes of thought the example of mathematics comes easily to mind as fitting. One might even speculate that if any cognitive activity requires special or modular organization these would. Let's look at simple mathematics. You can go just about anywhere in the world and you will likely find the Arabic system of numerals used to represent the concept of number, no matter what language is written and spoken. So you might expect that students performing arithmetic or computational tasks would engage the same areas of the brain. Thus, students who live in Great Britain or in China, when presented with the simple problem, what is 8 + 5? would write or say "13," with equal ease and speed. Are the underlying brain structures utilized to perform this simple math identical? The spoken and written languages are clearly different, but Arabic numerals that are the basis for calculation is the same in both countries. Is there a universal brain module for simple math?

Linguists, neuroscientists, and educators alike were surprised by the findings that strongly implied that "different cultural experiences can induce variation in brain function during mathematical problem solving" (Cantion and Brannon 2006). The study, done by Tang et al. (2006) applied brain imaging (fMRI) to native English and native Chinese speakers to measure the level and location of brain activity during the performance of four kinds of mathematical tasks. Three of the tasks involved manipulating Arabic numerals and the fourth tested the ability of the subjects to differentiate among ideograms, or pictograms, that varied in structure but were otherwise meaningless. What Tang and Co. found was that all four tasks engaged a common brain region, the occipitoparietal network, which reflects the use of visual pathways and an association area in the brain. That is not surprising. What was surprising was that arithmetic-induced activity in the brains of English speakers reveal greater activity in the left perisylvian areas than Chinese speakers. The left perisylvian areas are associated with language processing across cultures and languages. In solving the same arithmetic problems, Chinese speakers exhibited greater activity in a different brain area known as the premotor cortex. These experiments demonstrate a cultural dissociation for numerical tasks. What does this mean? Tang et al. "hypothesize that number representation and arithmetic processing in the brain may be affected by a variety of cultural factors such as educational systems and mathematics learning strategies that are beyond language-related experiences." In essence, their findings are consistent with the view that simple arithmetic seems more dependent on language processing than more demanding tasks, such as comparing sets of numbers and determining which is the larger or smaller. It is thought that this task increases the mathematical "loading" and requires different pathways in the brain (through the parietal-occipital cortex) in native Chinese speakers than in native English speakers (through the temporal cortex).

East and West: Other Cognitive Differences?

The study by Tang et al. raises the question of whether or not other aspects of cognitive processing differ between citizens of East Asia (Korea, Japan, and China, e.g.) and the West (U.S., Great Britain, Australia). A group of cultural psychologists led by Richard Nisbett and Kaiping Peng argues that cultural background or worldview strongly influences the style and mode of cognition. First, they point out that individuals from the Western world tend to see themselves as operating as "free agents" separate from the natural world whereas individuals from the Far East tend to view themselves as part of an interconnected and interdependent "web" of relationships encompassing society, culture, and physical environment. Second, how we categorize the world is sensitive to culture. When subjects analyze a photograph of tourists at the Grand Canyon, Westerners tend to focus on particular objects which get isolated from their context, both literally and figuratively, whereas Easterners pay more attention to context. In language terms, according to Nisbett, this means that in the West, nouns are emphasized; in the East, verbs. This culture-centric view means that cognitive differences, and presumably the brain mechanisms that underlie them are inseparable from socially and culturally induced ones.

A third example from brain scanning studies deals with a specific behavioral disorder, dyslexia, a language disability that affects reading and presents in both the receptive and the expressive sense. Dyslexia is a developmental disorder that affects 5–10% of the population. I chose this example because one might think that a disorder as specific as one that affects reading might arise from breakdowns in specific brain mechanisms—a reading module. A study reported by Paulesu and colleagues in 2001 in the journal *Science* compared English-speaking dyslexics and Italian dyslexics. Their analyses of comparative fMRI data led them to conclude that the brain areas affected were the same, independent of native language. They argued that for dyslexia, nature trumps nurture. However, in 2004, Siok et al. studied Chinese dyslexics and they conclude just the opposite—that the basis for dyslexia is not universal in the brain but depends on language community or culture-that nurture trumps nature.

What Is the "Take Home Message?"

In the brain sciences, there is a tendency to follow reductionism too narrowly and we need to be on guard against overly strict biological determinism. The concept of specific computational modules in the brain is an alluring notion but as our colleagues in psychology know well, it is very hard to disentangle nature from nurture. If society is looking to brain science to do the heavy parsing, early results caution: "think again." Certainly, the program to map behavior onto brain mechanisms is still in its early days and it is too soon to draw firm conclusions. The main tools for "looking into the human brain as it thinks" while breathtaking are nowhere near the level of refinement that neuroscientists hope to take the technology. The most advanced instrument, fMRI, produces results that are not without interpretational pitfalls. After all, what fMRI actually measures is the volume of blood-flow through cerebral arteries, as it waxes and wanes with metabolic demand, not the actual bioelectric events that are the currency of informational exchange in neural systems. So the bio–reductionist and determinist program of neuroscience may have only just begun and the best yet to come. But the findings of the cultural psychologists like Nisbett and others reminds us that arguments about the relative roles of nature and nurture are alive and well even in cognitive neuroscience.

References

Cantion, J. F., & Brannon, E. M. (2006). Shared system for ordering small and large numbers in monkeys and humans. *Psychological Science, 17,* 401–406.

Paulesu, E., Demonet, J. F., Fazio, F., McCrory, E., et al. (2001). Dyslexia: cultural diversity and biological unity. *Science, 291*(5511), 2165–2167.

Siok, W., Perfetti, C. A., Jin, Z., & Tan, L. H. (2004). Biological abnormality of impaired reading is constrained by culture. *Nature, 431,* 71–76.

Tang, Y., Zhang, W., Chen, K., Feng, S., Ji, Y., Shen, J., et al. (2006). Arithmentic processing in the brain shaped by cultures. *PNAS, 103*(28), 10775–10780.

A Task in Education: Towards a Community of Difference for *Learning to Live Together*

Chae-chun Gim

Why *Learning to Live Together* Again?

In 2016, the world was stunned by two unforeseen events. In June, the UK voted to leave the European Union, and in November, Donald Trump was elected as the president of the United States. The polls then showed that the chances of voting against Brexit were high, and the chances of Trump being elected were unlikely. The results were unexpected. The British people voted to break away from the EU and the Americans elected President Trump.

Why did these two events in 2016 come to us as a shock? Because the UK and the USA, leading countries in advancing human rights and democracy for a significant period of time in the modern era, seem to shift its stance from living together at the global level to living together at the national level. The UK and US have led the world in the direction of living together, well-considered as watch dogs of world democracy by expanding and advocating policies for free trade across borders, woman's rights, human rights of various marginalized groups, and aid for underdeveloped and developing countries. But now, many people are concerned that the UK's Brexit policies and US President Trump will focus on protecting its own citizens and erecting an impassable barrier at the border, both opposing policy directions from a world citizen's value of living together. These circumstances give us a reason to rethink the meaning of 'learning to live together'.

As the UK and USA have changed directions in foreign policy, the significance of the global agenda is further amplified. Climate change and resource depletion is a growing crisis, and the gap between the rich and the poor within and between countries is another emerging global agenda. The recently frequent terror attacks and immigration are more of a global issue than of individual countries. 'Learning to live together' can be one of the strategies to solve global issues under the current

C. Gim (✉)
Korean Educational Development Institute (KEDI), Jincheon County, South Korea
e-mail: ccgim@kedi.re.kr

© Springer International Publishing AG, part of Springer Nature 2019
J. A. S. Kelso (ed.), *Learning To Live Together: Promoting Social Harmony*,
https://doi.org/10.1007/978-3-319-90659-1_18

circumstances. Despite this possibility, the UK and the USA have shifted their stance away from cooperatively addressing the global agenda and solving international problems to focusing on their own political issues. This gives us another reason to rethink the meaning of 'learning to live together'.

In reflection of the past and in preparation for the future society in the 21st century, UNESCO created an education committee in 1993 and presented the 1996 Delors report 'Learning: the Treasure Within'. In this report, 'learning to live together' was suggested as one of the four pillars of education. This report thereafter significantly influenced international organizations such as the UN, UNESCO, and many countries worldwide when establishing the direction of its education. Despite such international efforts, why, then, do the global issues aforementioned still exist? Why is the UK and USA striving towards learning to live together within the confines of their own borders rather than globally? More than 20 years passed since the Delors report argued for learning to live together, but why do global issues such as climate change, resource depletion, inequality between the rich and the poor, terrorism, and immigration remain unresolved? (Or more specifically, why is it getting worse?) Can education contribute to solving these problems? Awareness of these issues serves as the basis of this paper.

Education, a World-Changing Force in a World Driven by Economy

No one would deny that the economy greatly impacts one's life. The importance of the economy in our industrialized lives is indisputable. Economic inequality will worsen in the future as many jobs are replaced with automated machines and artificial intelligent robots, and lives will more heavily be impacted by the economy. The influence and significance of an economy on one's life was emphasized 2,300 years ago in a Chinese classic by philosopher Mencius. An excerpt is as follows:

> If the average citizen does not have a fixed sense of livelihood, then they cannot maintain a stable state of mind. Without a stable mind state, a debaucherous, cunning, and prodigious life all becomes possible.[1]

Mencius mentions here that if there is no fixed livelihood, there also is no stable state of mind. In other words, if the average citizen does not have a secure job, it becomes impossible to live an upright, moral life.

Politics seems to be another factor that greatly affects one's life. The UK Brexit and USA election of Trump resulted from civic participation through voting. I think

---

[1]The fixed livelihood refers to the average citizens being able to make a living from their jobs, and the constant mind refers to the ethics of an upright mind and behavior.

that this political practice was influenced by the state of economy in the UK and USA. The world crisis we face today is arguably caused by economic instability.

The economy and politics are main factors in causing global crisis but are also possibilities that can lead to solutions. We are, then, left with the fundamental question what the role of education is in creating a world where we can live together. Can education have an impact on the Brexit policies in the UK and the USA presidential election? The responses are likely to be negative considering that it is difficult for education to solve global issues in a direct and timely manner.

However, we place our hopes in education. Mencius argued that the principle of 'if there is no fixed livelihood, there also is no stable state of mind' is applicable only to the average citizen. The king, noblemen, and elite of society were expected to remain morally upright regardless of economic instabilities in life. This is where we can pinpoint how education can contribute to creating a better world where we live together. Unlike the time of Mencius in the 4th century B.C., we can now teach in the 21st century to both the societal elite and the average citizen to live an upright life. Within this context we can identify meaning behind our educational efforts for 'learning to live together'.

What Is Education for?

What is the purpose of school education?[2] This question does not refer to the purpose of school education in the past when only a minority of students attended schools.

It does refer to what schools are for in the public education system that began in the 18-19th century. This question is introduced to probe into the reason for the public education system. Why did the public education system emerge in human history? Let us take a look at the case of the USA where there were societal debates during the process of creating a public education system. There was the common school movement in the 19th century. The advocates and sponsors of the common school movement, which created the public education system for grades 1–12, were made up of the clergy, politicians, and factory owners; the main purposes of the movement were political, cultural, and economic. From the late 18th century to early 19th century, the rapid increase of immigration to the US raised the need for a structure to uphold the White, Anglo-Saxon Protestant (WASP) culture as a dominant social and cultural framework. The labor force demanded people with basic cognitive abilities of the 3Rs as the 19th century became increasingly industrialized and urbanized. In summary, the purposes of school education in the

---

[2]To be specific, there is a difference between education and schooling. Education is a general term referring to the growth of an individual through learning, while schooling indicates the kind of education that occurs in a specific school setting. Schooling can be viewed as a sub-division of education. The terms education and schooling used here, however, are not strictly exclusive but used interchangeably.

public education system included both political, cultural purposes to foster citizens identifying with WASP culture, and building a labor force equipped with basic skills needed in factories.

Although the public education system during the 18–19th century in each European country appears slightly different, each country's public education system evidently held political, cultural, and economic purposes. As the Imperial Age in the late 19th century began, the public education system in Europe and the USA expanded into many countries around the world. School education in each country included the purposes of creating citizens with a national identity and active citizens for the economy. It can be said that the international community began to speculate the purpose of school education in the early 1970s. After riots erupted around the world among the youth including college students in the late 1960s, the UNESCO Commission presented the 'Learning to Be' report in 1972. This report emphasized maximizing the use of students' individual talents while also maintaining the purpose of school education to raise students as main agents for the economy as was conducted in the 19th century industrialized society. A significant point of this report was that education was no longer premised as transmission of knowledge but as a means for students to secure their future and careers and to actively participate in creating the society of tomorrow.

In 1996, UNESCO presented the Delors report titled "Learning: the Treasure Within.[3]" In the early 1990s, the failure of worldwide progress, job production, and social integration policies required a reexamination of educational goals. The following four pillars of an effective education system summarized in the Delors report is the product of work beginning at 1993: learning to know, learning to do, learning to be, and learning to live together.

The UN and UNESCO recently presented declarations regarding education. In 2015, the UN declared Sustainable Development Goals (SDGs) in "Transforming Our World: the 2030 Agenda for Sustainable Development." The SDGs consist of 17 goals including ending poverty and hunger, improving health and education, making cities more sustainable, combating climate change, and protecting oceans and forests. In 2015, UNESCO presented the Incheon Declaration titled "Education 2030: Towards inclusive and equitable quality education and lifelong learning for all" at the World Education Forum held in South Korea. Education for Sustainable Development (ESD) and Global Citizenship EDducation (GCED) were adopted as one of the main agendas of the declaration and included as a part of SDGs. Even though the ESD and GCED are closely related to 'learning to live together', it remains to be seen whether it will contribute to the solution of the global issues we are facing today.

---

[3]In 1997, the OECD presented the "Definition and Selection of Key Competencies" as a product of the DeSeco Project. The three competencies suggested by the OECD are the following: (1) use tools interactively, (2) interact in heterogeneous groups, and (3) act autonomously.

The four pillars mentioned in the Delors report in 1996 can be divided into two categories.[4] Attaining and utilizing knowledge at the individual level are the first two pillars while the last two pillars go beyond the personal to the level of relating to others and society. To put it differently, education has the task of developing ethical minds and social skills as well as that of attaining and utilizing knowledge.

Although school education successfully advanced the attainment and use of knowledge as an economic agent, it failed to address ethical aspects such as intra- and inter-personal skills. What type of people must education raise? Is there a way to grow moral individuals who well understand themselves and others, and learn to respect and live peacefully in the midst of differences and diversity? Is it possible to raise agents who can grasp the seriousness of global issues such as climate change, resource depletion, and inequality and strive to bring solutions to these problems?

Education for Raising Ethical Agents

We understood the first two pillars of the Delors report as the means to grow students into agents for the economy and the last two for ethical agents. For the sake of discussion convenience, let's call the first as 'education for employment' and the latter as 'education for ethical (thinking-capacity) development'. As the Delors report points out, both employment and ethical development are important factors in education; we cannot adopt one and abandon the other when implementing education. We need to strike a balance between the two.

Starting from the late 1990s, however, unemployment and job instability among the young adults in their 20s worsened with the advancement of automated machines and artificial intelligent robots. Employment was not guaranteed even after earning a college degree, and when employed, they faced unstable jobs and low income. This generation is referred to as the 880,000-KRW-generation in Korea and as the 700-Euro generation or 1000-Euro-generation in Europe.

Unstable labor market conditions pushed school education to place more weight on 'education for employment'. Schools focused on raising the capacity of students needed in the workplace, and consequently, education for ethical development diminished in schools. Competitiveness was a keyword emphasized in schools, and terms such as respect for differences and sustainable development were hard to find in schools.

However, the world we are living in today is facing rapid changes. Global crisis such as climate change, resource depletion, inequality between the rich and poor,

---

[4]There can be objection to the interpretation of the first two of the four pillars in the Delors report for individual attainment and utilization of knowledge, and the last two for character, social and citizenship skills. Various interpretations of the Delors report four pillars are possible. However, the presenter views that learning to be and learning to live together have been relatively neglected in school education so far, and willingly interprets education as a method to raise moral agents by directly teaching about the meaning of human existence and social/citizenship skills.

terrorism, and immigration is ever increasing and many jobs are disappearing with the development of the internet and artificial intelligent robots. If fundamental measures are not taken soon, as time passes, these problems will only worsen and remain unresolved.

Let's try to predict the results of each country's movement in preparation of the 4th Industrial Revolution Age. Countries will most likely invest more money in elite education to ensure economic dominance in the 4th Industrial Revolution Age and take more interests in education that will help with employment. This will lead to greater inequality and income gap between the elite and the average person. As the number of jobs decreases worldwide, education geared towards the elite and employment will turn into an education in which winners take all or a zero sum game instead of education for all global citizens to live well together. Humans do not exist for the sole purpose of employment or making money. To live a fulfilling life, one must go beyond employment. One should develop capacities for democratic citizenship, enjoyment of culture, scholarly reasoning, and enjoyment of learning itself as kind of play. Based on these grounds we must be attentive to negative implications of an education revolving around employment and securing economic status.

Therefore, in order to provide solutions to the global problems we are facing, we must shift from 'education for employment' to 'education for ethical development'. In the future society regarded as the Fourth Industrial Revolution Age, capacities needed in the workplace (such as creativity, emotional intelligence, and social skills) are closely related to results of 'education for ethical development'. In this context, we need to take more interest in 'learning to be' and 'learning to live together' in future education.

Politics of Identity Versus Politics of Difference

Just considering 'learning to be' or 'learning to live together' suggested in the Delors Report leads to another question. This is because when education attempts to develop an ethical agent, we are left with the question of what an ethical agent will consist of. This requires a discussion that is complex, metaphysical and ethical.

Korea has a history of 5,000 years. In its long history, the country suffered many times for the invasions of foreign powers. Nevertheless, Korea has held a long tradition of a homogeneous nation. It had taken much pride in its homogeneity. When I was growing up, I was also taught to take pride in our homogeneous country. However, various groups of immigrants and foreign workers entered the country since 2000. Instead of an education emphasizing the identity of a homogenous nation, the Korean education now emphasizes acknowledgment and respect for differences and multicultural education.

The USA is known as a country of immigrants of different origins all over the world; however, instead of acknowledging and respecting immigrants as they are, the USA pushed for the adoption of WASP culture. This preference was so strong that the Latin phrase *e pluribus unum* (out of many, one) was engraved and printed on the penny and one dollar bill. The attempt to have all immigrants adapt to the WASP culture was reflected in the melting-pot politics. The public education system took on the role of putting all so called 'impurities', differences, and diversity of immigrants into a melting pot to mold a distinct American identity. Korea was once a country that attempted to preserve a homogeneous nation and identity, and the USA attempted to advance a single WASP identity despite being a country made up of numerous immigrants from all parts of the world.

It becomes difficult to acknowledge differences and diversity when a nation or society enforces a single identity. Those who identify themselves differently from the dominant culture are faced with discrimination rather than acceptance and respect for their differences. The discrimination that foreigners faced in Korea and non-whites in the USA are all results of stressing a sole identity. In order to prevent differences from degenerating into forms of discrimination, we must uphold a positive attitude towards differences rather than emphasizing a single identity. Furthermore, respect for differences and diversity is in itself significant, and also holds the possibility for learning, growth, and creating something new through the encounter with differences or diversity.

Appreciation of differences and diversity in countries such as Korea and the USA is important and, needless to say, it is more important at the global level. Respect for differences and diversity has become especially more important as our world is interconnected through rapid transportation and ICT. The basis of living as global citizens requires respect for differences of race, ethnicity, language, culture, religion, etc.

When we speak of raising ethical agents, what kind of ethical agents will we need? They are individuals who have a respect for humanity and nature. The ethical citizens we need are those who take interest in sustainable development of humanity and nature, display awareness and respect for differences and change their own personal identity to prevent perpetuation of discrimination of those different from themselves. They also must embrace differences and diversity as a source of learning and growth. They must deconstruct unceasingly their own identity as they encounter others and create a new and larger identity, similar to an ethical agent like Emmanuel Levinas.

Levinas and the Moral Agent

Emmanuel Levinas is a philosopher who taught the importance of the other, hospitality, and welcoming the Other. To understand Levinas as a philosopher who proposed ethics as the first philosophy, we must take a look into his personal history. Levinas was born in 1906 into a Jewish family in Lithuania, growing up well-accustomed to the Jewish and Russian culture. At the age of 17, he attended the University of Strasbourg in France and was captivated by the French philosophy and culture. While at Freiburg University in Germany, he studied phenomenology under Husserl and Heidegger. During the Second World War, he suffered the loss of his family at Auschwitz. By looking at the context of Levinas's life can we understand his ethical argument that we must embrace the sufferings of the 'widow, orphan, and sojourner'.

I am personally interested in Levinas's concept of growth through evasion and transcendence. Levinas defined the 'relation of the ego to the self' as ontology and the 'the relation of the ego to the Other' as metaphysics. Levinas proposed to go beyond ontology unto metaphysics, in other words, going beyond a relationship to one's self to a relationship with others. Here, Levinas highlights a forming of a new agent by making a relationship with another.

Levinas premises the ethical agent within a relationship with others. For the agent, the Other is an infinite being that the agent cannot fully grasp or understand, thus leaving the agent incomplete in forming a new self while relating to another. Inevitably the agent will recreate itself every moment of engaging and relating to the Other. Within the relationship with the Other who possesses infinite qualities, the process of constructing and deconstructing the agent is endless. In sum, Levinas argues that the growth of the ethical agent occurs when one goes beyond the self into accepting the Other to form and engage in a relationship.

Suggestions for Change

It is difficult to prejudge the role of artificial intelligence in the future society regarded as the Fourth Industrial Revolution Age. There are predictions that by 2045 our generation will face a point or singularity in which we will lose control over artificial intelligence as it surpasses human ability. We cannot know when we will arrive at this point, but what we do know for certain is that the progress of automated machines and artificial intelligence is replacing existing jobs. New types of jobs are created from the development of technology, but they are not as many as the numbers of jobs that will disappear. Employment in the future seems hopeless and bleak as we witness the younger generation already struggling to find jobs.

Because of an unpromising future, schools tend to strengthen education that sets employment as the goal. Education for raising ethical agents is weakening, and in the case that it is taught, it ends with mere talk; the educational goal of raising economic agents is what actually exerts influence in our reality. There is even a tendency for schools to use 'learning to live together' as a means to raising economic agents by emphasizing cooperative learning in the context of education for employment.

I would like to conclude with three suggestions regarding education to create a world where we can live together. First, we need to make a paradigm change in education. Two issues are to be seriously engaged in the new paradigm: the first is a politics of difference, the second is global sustainable development. Education should not confine itself to raising economic agents but expand its premises to raising ethical agents who will embrace differences and actively respond to global issues. Rather than a community of exclusive solidarity surrounding a single identity, there is a need for a community of difference in which ethical agents constantly deconstruct their own identities while engaging with others in order to recreate new forms of identity.

Second, we need to teach SDGs at every school worldwide and ensure that the spirit of SDGs is embedded throughout the entire educational process of curriculum, teaching and learning, and assessment. Each school must instruct students in SDGs, whether by including them in existing related subjects or by creating a new subject entirely. Rather than teaching SDGs through textbooks, school education should provide students with a variety of opportunities to learn about global issues through both direct and indirect experience, and for each student to take small steps towards resolving these issues from where they are. Without the active efforts of the school and educators to teach the spirit and values of SDGs, they will remain as mere talk in educational literature.

Third, there is a need to resuscitate humanities in education. Needless to say, humanities have been the main content in education for a long time since the classical era because it provides a world view in which to consider the meaning of life and to act in accordance with his/her belief. Giving a renewed emphasis on humanities in education will be a good step towards raising ethical agents. As we observed Mencius's argument earlier, with education, even without a stable life, one can remain morally upright. It is at this point where we discover the power of education. Education should not stop at raising economic agents but proceed to raising ethical agents and also demonstrate the possibility that ethical agents can remain standing firm even as they go through failures as economic agents. How about organizing a new foundation consisting of humanity scholars interested in sustainable development and global citizenship education?

Healing: An Imagination Beyond Violent History with Focus on Korean Historical Context

Byong-Sun Kwak

Even in this civilized era, still people somewhere in the world are suffering from the consequence of violent historical causes. Korean history during the last century is, in a word, a history of "the wounded". Early in the 20th century, the nation had a painful wound due to the Japanese colonial régime. Furthermore, right after World War II, the nation was divided into South and North. It was a really contradictory and paradoxical decision for the nation already victimized by Japan. This artificial and unacceptable division has piled up historical pains not only for Koreans themselves, but also for the outer world. The Korean War during 1950–1953, a disastrous trace of the imposed division was a civil war in nature, but it became a concomitant international war, 16 countries dispatched armies under the UN flag fighting against the North Korea. Even though the war was suspended in 1953 by the armistice, the road to national unity is still blurred with its deepening inter-relatedness and complications from all sides. From a historical retrospective, a question is raised, "What has to be the driving will for agents interested in the future of Korea?" Among possible options, the foremost important stance is that the new history must be initiated to proceed beyond violence.

Whatever reason the status quo cannot be justified any more for its unrighteous and temporal trait. Answering this uneasy question, we require a "quantum jump" in our overall thinking, that is, a radical departure from the self-contained thinking to the open-mindedness centered on the dignity of whole human being. In this vein, thinking about healing in history is positively suggested as a way to overcome violent history. A main idea of healing is to honor the history of people whoever they are. Acknowledgment and reconciliation of the people involved in the history of pains and building a positive consciousness for harmony in the future can follow as due process. It is fortunate that there are stories about healing in Korea and New Zealand. Korea's case is a story about reconciliation at village scale between vil-

B.-S. Kwak (✉)
Incheon National University, Incheon, South Korea
e-mail: kwak@kedi.re.kr

© Springer International Publishing AG, part of Springer Nature 2019
J. A. S. Kelso (ed.), *Learning To Live Together: Promoting Social Harmony*,
https://doi.org/10.1007/978-3-319-90659-1_19

lagers having pains during the Korean war and New Zealand's case is a national scale about reconciliation between indigenous Maori people and immigrant white people, which signifies that history can be healed by heightened consciousness. Peace and war depend on the mind of people who are in charge in the given situation.

In regard to the agony of Korean people still ongoing, a promising solution for them to live together as one entity must be explored. The solution should ensure reconciliation of the two parts. Healing in history is an idealistic alternative in today's stern reality. However, it is not the unrealistic value only dormant in our mind. It should be a universal hope for all humankind. Through the spirit of healing, the unity of Korea can be made between the Korean people and outer world removing artificial division of the nation and between Koreans themselves rather than antagonizing each other with different political systems. This profound task needs joint effort from both the Korean people and the whole world due to its deeply interconnected nature and its geopolitical aspect. In this, vocabularies on healing in history must be tossed around and discussed as much as possible via agents, whoever they are, committed to making their own history with a new vision and notion for living together.

Our Human Condition with Remembered Past

Whoever we are is inseparable from our conception about the remembered past regarding to ourselves and to other people in relation to our existence. What we remember about our past constitutes an essential entity for the identification of ourselves. If we regard remembered past as history as Lukacs asserted (1985), history is certainly a determining factor in shaping what we are. The remembered past differs from people to people depending on collective entity such as ethnicity, language, religion, sovereign, value orientation etc. In relation to history some people may be proud of themselves for glorious passages of their former generations, some may lament for passive path of their ancestor or some may deplore suffering by oppression of unrighteous power. Our living condition here and now is directly intertwined with the remembered past whatever it is. In other words, the condition of life of a group of people in these days is shaped by the consequence of actions taken by agents in the realm of historical transit.

What we have as the remembered past is a fundamental asset for us to identify who we are, where we are from, what lesson drawn from it and what mission given to us for the future. That is why every nation has paid keen attention to recording of remembered past and teaching of it. Then what is the core of the remembered past as history? "What is history?" itself is an interesting theme with entailing controversial issues in developing relevant meaning. With different arguments and assertions on the nature of history, an overall observation can be drawn as follows:

Historical outlook has been evolved throughout a series of epochal changes toward deeper understanding of the human condition.
Remembered past is selective among human actions according to our conception and value orientation what is true and important.

Nowadays historical issue is easily converged as a global issue, even for fragmented local issues due to the inevitable interconnections of human life through deeper learning capacity and the renovation of information technology.

In historical deployment throughout millennia, historical figures and agents have transformed from legendary figures such as heroes, kings, lords, rulers to collective group of people actively participating in historical events.

Even in this civilized era of a wide spread of democratic ideal and advantage of high information technology, it is reality that there are still some contradictory and paradoxical problems remaining which require a quantum jump in our historical consciousness in order to overcome them.

We can assume that remembered past associates with our life in many ways. However, from an existential perspective any human being in the world from the beginning to until now has not been destined to live in a particular condition within such factors as ethnicity, gender, social status, geological environment etc. programmed prior to the birth. Belonging to a community of certain traits cannot be made by our own choice; rather it is given as it is regardless of our own will. In this connection, it is appropriate and legitimate to think that individual persons whoever they are, are born with dignity and without discrimination. Of course, the mandated thinking in this kind only appeared three hundred years ago through the Declaration of Independence of the United States and lastly manifested in 1948 in the UN Declaration of Universal Human Rights which has been institutional form and guiding principle as global norm for the whole human being. Even with this evolved consciousness on human dignity, it is reality that there are regions in the world where people are under threat of violence by organized group or nation venturing for their interest at all cost which eventually cause human suffering in a vicious circle. In reality there is violence latent everywhere due to the never ending conflict and crisis over selfish interest such as vested interest, sovereignty, territory, hegemony etc.

However, there are some positive moves even limited scope that we human beings can transform the history of violence into the history of healing. In recent years, I found two cases of healing, one in Korea and the other one in New Zealand, which have motivated me to boost up healing as a way of thinking in dealing with conflict related to historical background. Above all the foremost motivation for me being interested in healing history is what has happened to me throughout my whole life so far which is inseparable from the reality of history. I was born four years before the end of the World War II as a son of a Korean immigrant in Mudanjiang, a northern part of China. At the time the region was ruled by puppet regime of Japanese colonial power. Right after the world war, my family immigrated back to South Korea. From this early child life until now my whole life has been with the panoramic historical fluctuation through the independence of Republic of Korea, Korean war, so called Korean miracle economic development from ashes of the war, democratization of the nation from authoritative structure. The experiences that my generation have during the last three quarters century in South Korea in particular are so dynamic and compressed that there might be no such blessed single generation in the world before who could have experienced such variety. It has taken only several decades for Industrialization and democratization to achieve what is known to take several hundred years in the western world. I watched the possibility of collective will for progress in the meantime, I saw the violence of history with my own eyes and also the modesty of human mind which enables us to move forward living together. As an educationist, I am always dreaming my homeland the South and North Korea unite as one nation living together. In this historical task, healing in our history can be an ultimate hope for the solution of human trouble not only for Korea but also anywhere in the rest of world.

History of Paradox and its Violent Nature

In this civilized era, still people somewhere in the world are suffering from the consequence of violent historical cause. Korean people is one such case. Korean history during the last century is a history of the wounded. Early in the 20th century, the nation had a painful wound due to the domination of Japanese colonial regime. In its whole history about five thousand years, Korea has been a peace-loving country which has never invaded neighbors while being attacked by them countless times. The ideal of Korea in its historical origin is the world of human being with maximum benefit for all. All this historical legacy along with other identical traits was painfully damaged by the Japanese colonialism.

Furthermore, right after the World War II, the nation was totally alienated and pushed being divided into the South and North part by the Allied U.S and USSR as a resolution of post- World War II in 1945. The dominating agents of historical change at that time were ignorant about the yearning of Korean people for independent nation as the due course of struggle against Japanese colonial regime. It was really a paradoxical decision for Korea which had already been victimized by Japan. This artificial and unacceptable division has made Korean people strive for both national independence and unification as inevitable tasks, through which unnecessary historical pains have been piled up not only for Korean themselves, but also for the outer world.

The Korean War during 1950–1953 was a disastrous trace of the imposed division. It was a civil war in nature, but it became a concomitant international war, 16 countries dispatched armies under the United Nations flag fighting against North Korea. With the armistice in 1953 the war has been suspended. Since then, several attempts for mutual cooperation and unification have been tried from the both South and North sides but in vain. On the contrary, the road to national unification is still blurred with its deepening inter-relatedness and complications of division issues from all sides. Retrospective of this history, a question is raised, "What would be the ultimate proposal for all agents interested in the future of Korea to break through the entangled status quo situation?" Among possible options, the foremost important stance is that the new history must be initiated to proceed beyond violence.

If men and women as the agents of history are detached from the decision for their own history, then the coming history happens to be contradictory and violent as it was. As much as we are being alert and paying a keen attention to the choices of leading agents, we are able to affect history in a more positive and responsible direction. However violent it was in the past, in a broad sense, it was the result of human mind involved in its time and space. What happens is almost always involved with what people think happens (Lokacs 1985, p. 137).

Cases of Healing: Village Community Scale in Korea and National Scale in New Zealand

Whatever reason why the current historical context as it is, the status quo cannot be justified any more for its agonizing pains and thereby its temporal nature for Korean people. Answering this uneasy question, I think we require a "quantum jump" in our thinking about possible choices among many alternatives, that is, a radical departure from the self-contained thinking centered on vested interest in every sector to an open-mindedness centered on the unity of Korean people as a whole. In this vein, thinking about healing in history is positively suggested as a way to overcome violent history. Main idea of healing is caring for each other for coexistence and living together. In practice, such processes as acknowledgment of remembered past in honor and respect, reconciliation of the people concerned and building a positive consciousness to live together in harmony in the future are suggested to be in consideration. It is fortunate that there are stories of healing at local village scale between people antagonistic to each other caused by suffering during the Korean War in Korea and at national scale between indigenous Maori people and immigrant white people in New Zealand, which signifies that history can be healed by heightened consciousness. Peace and war depend on the mind of people who are in charge in the given situation.

Reconciliation and Healing at Dado Myeon in South Korea

Dado Myeon is a mountainous village located at outskirt of Najoo City, a middle city in southern part of South Korea. In 1948 South Korea was established as the legitimate government under the recognition of the United Nations. When the Korean War broke out in 1950 by the invasion of North Korea, the security of this area became unstable due to the rise of pro-communist partisan guerrilla activity. Armed clashes between government forces and the guerrillas was inevitable and the struggle continued for a while even after the armistice in 1953. A series of fierce battles claimed several hundred civilian casualties, who were innocent villagers living together in a peaceful community. They were victimized either by the government force or by the partisan guerrillas. There were some people actively engaged in left wing camp against the government but majority of the village were indifferent to political ideology. History was really violent for these remote villagers. This tragedy has made the villagers divide into two sides between offenders and victims from reciprocal stance with bitter resentment. Since then for several decades the bereaved family of each side held memorial services separately for the deceased.

In the meantime, the surviving families of the village voluntarily organized a committee to arrange a joint annual memorial service for all victims regardless of the cause of tragedy in the spirit of reconciliation and healing in the mid of 2000s. Along with this voluntary action, the government launched a national committee for truth and reconciliation in order to identify what really happened for salient cases of tragedy during the period between the liberation and the end of Korean War. With this joint effort from the government, the descendants of survived family acknowledge that the painful accident was due to the lack of understanding about the overall circumstances at the time and eventually the deceased all were the victims of violent history regardless of whether they were in government side or sympathetic with guerrilla. At last the survived families reached common under-standing that each side forgive each other reciprocally for the past of ancestor's wound and move forward for reconciliation and unity of whole village. It is a great step for villagers of historical pains to overcome mistrust and antagonism incurred during Korean war. Movement of this kind echoed to another neighbor village called Gurim about 50 km away from the Dado Myeon. The Gurim village suffered the same wounds as Dado Myeon.

Healing at National Scale in New Zealand

The case in New Zealand is about a story of healing between indigenous Maori people and immigrant white people, which signifies that history can be healed by heightened consciousness. The case of New Zealand is well reported by R. Consedine and J. Consedine, in their book titled "Healing Our History"(2012). The book shows how the immigrant white people of colonial domination and the indigenous Maori people who were humiliated and suffered under colonial rule in the past can unite as one people of New Zealand. The road to healing in New Zealand is an ongoing process, meaning that there are issues to be settled through deeper mutual understanding and respect for indigenous culture in order to over-come racial discrimination and inequalities. It is noted that this unprecedented collective effort has been made by people in an advantageous position with a new concept of "healing".

> History cannot be changed, but sometimes it can be healed. History must be healed if humankind is to survive (Consedine and Consedine 2012, pp. 298–299)

This movement began in the mid-1980s with the emerging demand for non-discriminating access to resources and decision-making processes in social organizations and for respect for indigenous culture on the part of the Maori people and with the changing conception that nothing will change until historical grievances are recognized and confronted (Consedine and Consedine 2012, p. 300). The dominating white people (Pakeha) with some fears showed courage to recognize the existence of Maori culture in its structure and to moving forward as a bi-cultural society.

The case in New Zealand shows that healing is possible where people in the position of privilege acknowledge the pain of underprivileged people oppressed by the former while the latter is demanding their rights guaranteed by laws and by the universal spirit of human dignity. In the case of New Zealand the concept of healing is applied as a powerful way for both dominant white people and wounded indigenous Maori people through the so-called treaty education process which addresses mutual comprehension and respect, and acknowledges painful history, forgiveness, reparation and restoration of right relationships, and reconciliation. The immigrant colonial white majority acknowledges the indigenous Maori culture as it is and develops re-conciliated identity through the full partnership between the two peoples.

Healing as an Emerging Historical Consciousness

The story of healing at a local village scale in Korea and ethnicity reconciliation between colonial dominating white people and indigenous Maori people at national and cultural scale in New Zealand is a small step but a quantum jump in our total history that deserves keen attention from the people of the world. These are concrete examples showing the great potential that whatever differences they have, people can develop living together in this narrowing globe day by day. The Korean case is a story of reconciliation of a village where the habitants had fatal pains due to violent historical causes that happened about 70 years ago in a situation where the whole national security was in crisis under war. It is a case of healing at micro level, however it sheds a significant light for Korean people enabling a departure from the short-sighted view centered on fragmented and vested interest into a encompassing long range view in order to live altogether in reconciliation and harmony. This conceptual transformation must be encouraged and matured for the whole of Korea including the North and the South becoming toward one nation.

The civilian casualties at Dado Myeon during the Korean war were only victims of violent historical causes which were imposed from outer circumstances at the time. They were generous neighboring villagers sharing each other as it was in traditional rural life. All of a sudden, some of them were killed by the left camp guerrilla for the reason of not helping them, and some by the police for the reason of helping the guerrillas. This happening was mainly due to the division of the nation by the Allied powers. The imposed division brought truly appalling consequences including not only ideological division of people between the pro and con for free democratic society and vice versa for communism–but also the military attack of North Korea on the South.

In colonial era, the pain was caused by the domination of alien power which sought exploitation as much as possible from the ruled country. It was historical irony that after the liberation from the colonial regime the pain incurred from the inner conflict was severe and destructive for the self-esteem of national pride. Even though more than seven decades have passed since the liberation, still there is no positive sign observed for mutual cooperation and partnership for the common goal of national unification. Rather tension in the Korean peninsula is becoming serious for the whole world. However, it is presumed that some day sooner or later they will come closer to the table of dialog for mutual respect and coexistence. In an overall view, the provocative action of North Korea may be mainly caused by the artificial division of the country seventy-odd years after World War II.

Perhaps, the day that each side gets together for mutual comprehension and respect toward building a unity of whole Korea will be possible when the agents related to the Korean dilemma can confront the historical violence as it was, acknowledge bitter pains and wounds of people through the reckless violence of history, and seek a peaceful way through laying down the vested interest of each fragmented part. This implies that the agents concerned with Korean situation have to have a common ideal that no more violence in the Korean peninsula is allowed.

In this vein, healing in history can be an ultimate alternative not only for Korean dilemma but also for human beings as a whole moving forward to living together beyond violent history. Whether healing can be a future historical consciousness depends on our awakening to what the highest value is for the future of human beings. Peace and war depend on the mind of people who are in charge in the given situation. If healing can be a mainstream of historical consciousness in historical thinking, the leading agents in historical time and space care about the reconciliation and harmony rather than exercise dominating power over the disadvantaged for selfish interest. Selfish interest includes domination, hegemony, and maintaining status quo of exploitation for vested interest.

Healing can occur at many different levels in historical settings. At micro level healing can be possible between families in feud. The Dado Myeon case is at a village level intentionally taking place in collaboration with the government's supportive measure. The New Zealand case is at a national and racial level as well. The Dado Myeon and New Zealand cases are historical actions taking place in a country within its intra-national structure. Beyond this within national structure, healing can be applied in a wider inter-level context—between nations, inter-racial differences, inter-cultural differences, inter-ideological differences, inter-religious orientations and so on.

Whatever context or whatever level healing is designed, such processes as exploring the causes of historical violence, acknowledging the pain of history, offering genuine apology, and making reparation are presumed to be taken into consideration as important elements for developing the common goal of mutual understanding and cooperation for all agents concerned. Healing history may not be possible for those who think the human race is doomed to a never-ending cycle of war and violence (Consedine and Consedine 2012, p. 299). It is only possible for those who believe that human beings can change their way of thinking concentrating on human dignity regardless of their origin. In the past, human life was scattered and fragmented mainly along the dimension of tribal origins and geographical conditions. Divided groups have competed against each other not only for survival but also for domination and hegemony. In this group competition for survival, Greene (2013) reports that groups with wider cooperative capacity continuously have survived in history. It means that the groups who have sought interests of wider community beyond the demarcation line of selfish ones gain more benefit for the community as a whole. In this context, healing can be a supreme value in which we all can aspire for a cooperative world.

Healing our history can definitely be a starting point for all contemporary humanity toward the world of coexistence. In order for historical agents to proceed to the healing stage is a formidable task: the emphasis and choice of actions can be different depending on existing values and context. In general, such ideas as joint research in humanity centered on history, writing common historical textbooks between or among nations which were previously conflicting over history, continual confirmation of human dignity, rejection of violence and war, respect of identities, honoring of history of people disadvantaged in the past, etc., are worthy to consider.

Based on this assumption, the start of healing process begins recognizing what pains and damage happened in the past history "Proper recognition of wrongdoing paves the way for the beginnings of transformation, setting the scene for further steps toward reconciliation and healing (p. 301)". Genuine apology follows acknowledgment. Forgiveness goes along with apology. The practical way of acknowledgment, apology, and reparation can differ from contextual circumstance to circumstance.

Along this line, in order to overcome Korean dilemma a healing process at macro level can be imagined in which the agents who have made division of Korea 1945 acknowledge that it was really a wrong choice for Korean people, honoring the wounds and pains of Korean people incurred by that decision, and explore joint effort for the two Koreas as one nation. This transformation cannot be made by one or two nations concerned. Perhaps a declaration of healing by international organization like UN as a total process not only for the whole Korea, but also for the rest of the world wherever the code of universal human rights is in crisis.

> Nonetheless, we continue to live our lives in the tragic gap—tragic not simply because it is heartbreaking but because, in the classical sense of tragic, it is an eternal and inescapable feature of the human condition. This is the place called human history where we must stand and act with hope even though neither we nor any of our descendants will see the gap permanently closed (Parker 2011, 191–192).

> We believe that the good will which was given to our ancestors from the heaven and had been kept in the bottom of their heart and passed over to us throughout a long historical journey, if we commit ourselves on it, must play the central role in the mission of future of world history (Ham 1965, p. 437 translated by author).

References

Consedine, R., & Consedine, J. (2012). *Healing our history*. Penguin Books.
Greene, J. (2013). *Moral tribes*. London: Atlantic Books.
Ham, S. H. (1965). *Korean history in its meaning*. Seoul: Samjungdang.

Hutchinson, F. P. (1996). *Educating beyond violent futures*. London: Routledge.

Lokacs, J. (1985). *Historical consciousness or remembered past*. New York: Schocken.

Kwak, B.-S. (2010). *Education toward healing of painful history*. Korean: Column in Webzine of Korean Social Science Research Association.

Palmer, P. J. (2011). *Healing the heart of democracy* (pp. 191–192). Wiley.

Ahimsa, the Way of Living in Peace and Harmony: Indian Perspective

SM Paul Khurana

Ahimsa—founded on the bedrock of amity, forgiveness, tolerance to foster global non-violence, peace and harmony—can be the way forward for humanity. Stern disciplinary actions and coercive approaches may help maintain law and order, but yet individuals may resort to violence and force to achieve their narrow aims more often than not, disrupting peace. For peace, we need to train individuals in groups in achieving Mahavira's idea of *Ahimsa Parmodharmah*, because it became supreme religion. This unique philosophy to life as such enabled Jainism in being recognized as one of the supreme religions in the first millennium that changed the moral climate of the Indian subcontinent. Ahimsa and peace are born in the mind of men but are achieved only by spiritual upliftment of individuals in a society. Like war begins in the mind of men, peace too emanates from inner consciousness. Ahimsa is one of the most civilized approaches to social life. Buddha observed, "*Our life is shaped by our mind; we become what we think or believe.*" Our lives are fulfilled if we are guided by compassion, love, kindness, joy in other's joy, equanimity, helping others in suffering. Ahimsa and Peace can be neither conquered nor imposed. It provides freedom, harmony and justice. It is not a political affair but brought by conversion of hearts. We do need to pursue the justice of god and believe in forgiveness. In our pluri-religious world the peace is not a myth but real and must be one that transcends boundaries of individual religions, be truly secular and inclusive for both universal harmony and compelling justice so that we share and care for the earth. Taoism is marked by creating harmony in all aspects of living by alternatives to have peace, like flowing water does vanquish even hard and strong rocks and obstacles. Mahavira used it to great effect deeply understanding self and the conflict. Gandhi believed in *Ahimsa* of the strong and in not retaliating

S. P. Khurana (✉)
Amity University-Haryana, Manesar, Gurgaon 122413, Haryana, India
e-mail: smpaulkhurana@gmail.com

© Springer International Publishing AG, part of Springer Nature 2019
J. A. S. Kelso (ed.), *Learning To Live Together: Promoting Social Harmony*,
https://doi.org/10.1007/978-3-319-90659-1_20

to violence with violence. In today's context of ongoing wars and the menace of terrorism that we face, it may not be unrealistic if we promote nonviolence ("except for self-defense.") or *Ahimsa*, which is the essence of all religions. Gandhi's opposition to war is total and *Ahimsa* would not fail but he agrees to killing, only to uphold the Dharma or one's duty if performed with detachment. Ishwarchand Vidyasagar explained Dharma as equal to virtue, righteousness, duty, quality, justice, religion, in that order. Vivekananda also asserted, "To *be pure (unselfish), to be good (doing good to others) as the religion*". Buddha said to his disciples: "Be a lamp to yourselves, seek refuge in you" and that we have no moral authority to kill any living being. If we are free from attachment and aversion alike, there comes the inner peace, concomitant with rising spiritually within us. Today we have many Gandhian peace activists in India, like Neelakanta Radhakrishnan, who have devoted their entire life to the cause of peace and non-violence and in restructuring Gandhi's Peace Army. Aspects relevant to Ahimsa are discussed below:

The following paragraphs are important in acquainting readers with the precepts of Indian philosophy as the background on the issue, to understand and appreciate Indian perspectives to the practice of Ahimsa. Violence was accepted in the past as it is now as an instrument of policy for toppling stable regimes. Buddha never waivered in his insistence on Ahimsa as the path to human salvation. Buddha remarked, "*They do not follow dharma who resort to violence to achieve their purpose*" (Buddha Dharmapada:xix, 256-57). Buddha saw that these destructive energies corrode the very soul and would turn the world into an inferno as violence produces permanent unrest, a cycle of assault and retaliation, resentment and rage. The antidote for all this was Dharma i.e. ethical conduct, wisdom and Ahimsa, Karuna and loving kindness. Indeed, one cannot build or win happiness over the unhappiness of others. In similar circumstances Gandhi would have insisted, "*Don't retaliate with violence*". As an agnostic Gandhian, Khushwant Singh, mentions that '*the only religious principle I subscribe to is Ahimsa or universal non-violence*'. Indeed it is Ahimsa that bears the essence of all religions, the rest is all marginal. When Osama-bin-laden and the Taliban were pitched against the USA, Khuswant Singh opined, "*Don't hurt anyone except when your own existence is threatened, but if we're to live rather than die, then something has to be done!*" It is 2600 years back that the Indian Saint Savior Mahavir while elaborating *Ahimsa Parmodharmah*' (non-violence) as the supreme precept of religion said '*A violent mind is not the place where peace would reside. In fact the world is not to be treated as the house of sorrow and pain but it is the door to peace and bliss which is ever open*'. Any sad and unhappy person has to first enter it, and then only the Bhava Parivartan (change in attitude) can help treat the underlying spiritual illness. This change brings divinity and the person attains peace and true bliss. It may be made clear that war begins in the minds of men and so does peace that emanates from the *chitta* (inner consciousness). It is not limited to hitting or harming others. We must get rid of dross or demeaning emotions, free of desires to rest in peace. A temple of peace can only be created on a stable foundation of *Ahimsa* as only a changed man can change the world or be a citizen of the world. Peace can neither be conquered nor imposed and it is this gift alone which is founded on justice and built in harmony. It is a non-political affair and can be achieved without fighting a war. Aurbindo made it clear that the problems of our existence are problems of lack of harmony. Let us understand that Parshurama represented a time in ancient India when social order was threatened by a power struggle between dominant segments of society, and the precept evolved that violence cannot be justified as a solution for resolving social problems. Gandhi does not equate Ahimsa with nonkilling. Having a cue from Lord Sri Krishna when he said to Arjuna (Bhagavad Gita iv;8) that He incarnates on earth in every age to restore *Dharma* by resetting all negative forces or having triumph over them. Yogi Krisna Prem* (1898–1965) was a Briton who underwent intensive study on Vaishnavite disciplines, and had the remarkable ability to express, in clear English, the depth and inner resonance of ancient Sanskrit texts. He concludes that—in darkness of prison of our body–our hearts, the birth of Sri Krishna, of truth, has to take place and win over kings like Kama, Krodha, Lobha, Moha, Mada and Ahankar. In Bhagavad Gita (xiv, 5) Krishna teaches Karma Yoga, which implies right action without attachment and that all things exist in ever varying combination of three qualities: sattva, rajas and tamas. Tamas is always much more with a small dosage of sattva while having a fair splash of rajas. It is not possible to fully eliminate tamas but to get a right balance for rajas and sattva is an objective that we should aim for.

Can Peace Prevail in the World?

The big question remains about peace: is it only a dream of the philosophers or poets or cranks? Jiddu Krishnamurti discussed the issue for five long decades and believed that nations do wage wars, but only to sign peace treaties. Perhaps more wars have been fought for religion than for love or brotherhood. Man's desires lead to sorrows in life in order to compete with others usually, for seeking pleasures, developing conflicts and a restless mind. Krishnamurthy concluded that problems created by the mind cannot be resolved by the mind alone. Nagarujana also observed that man's sufferings are because of his divided nature which can be balanced only by the void or *shunyata*.

Contemporary to Buddha, Lao Tzu of China who founded Taoism believed in the philosophy of living, to be detached from materials for benefiting others and harming none, leads one to the heavens. Confucius, however, diametrically opposite to Taoism, had a practical proposition, to have an ideal social order with high morality and righteousness. On the other hand Taoism does believe in untended, uncultivated field, in which all the growth occurs in right time/season on its own. In other words, a truly virtuous person is unaware and unconcerned of what others would think. It means everyone has virtues of humanity and righteousness. To lead a peaceful and fruitful life Lao Tzu's advice is to avoid competition, for the stress, unhappiness and conflict. Tao said all are having two beings within: a great one and another, the lesser one! He stressed on taking the hand of the great one, as it will lead us to greatness and the natural choice for us.

Ramakrishna Paramhansa (a devout Hindu Brahmin priest) was unique in advocating the positive acceptance of other faiths to avoid religious plurality. His experience was that all religions lead to the same goal, having said, "He.... who is called Krishna is also Shiva and the primordial Shakti, and he alone is called Jesus or Allah". Thus there is only one Rama and he has a thousand names.

Ramakrishna had beatific vision of Allah and also Jesus, as he had realized their identity with Kali, Rama, Radha and Krishna. He did show a new way to transcend religious barriers. His disciple, Vivekananda observed Practical *Advaitism* which means to all mankind as one's own soul based on the junction of the Hinduism (Vedantic) and Islamic theories to be the only hope for unity of two systems.

Tagore's Religion of Man

Tagore also suggests, based on his meditations on god, man and nature, especially on Vedantic perception that contact with the Infinite can only be through an independent exploration. He believed that an unknown principle was at work behind the entire creation of Almighty as he took a clue from the Bauls of Bengal, who sing of the divinity of man and love for all. Thus the infinite merged with the finite in humanity. And Tagore founded his religion: Religion of Man. He said that god is where the path maker is i.e. breaking stones and farmer tilling the hard ground. He felt that instead of blind beliefs, we should believe in the innermost truth, the life-sense of which is an extra awareness, greater than material sense. But the ancient percept was, to know thyself and Tagore felt, "Be known to others", there being no limit to this unfolding of inner self.

Comparison with Present Beliefs

We must not forget that we have to walk the forked path of life. Dalai Lama's personality epitomizes the dilemma of the forked path in everyone's life and an example of how shuttling between inner and outer lives, can be gracefully managed. Dalai Lama views the world as a mirage as taught by Buddha. The world has no permanent essence or reality, that it was a phantasmagoria of "images". Nagarjuna grossed it as a theory; the farther we are from the world, the more real it appears to us and the more closely we approach it, the less visible it becomes. Buddha suggests that our lives will be easily fulfilled if we are guided by compassion, loving kindness, and joy in other's joy, equanimity and attentiveness to other's sufferings. He reminds us that we do not act independently so much as we react to stimuli. Therefore, we must act based on wisdom and follow the ethical practice that Buddha had in his mind, when he said to his disciples, Be a lamp to yourself, seek refuge in yourselves."

To knock at the door of your neighbor's heart, you must forgive and sleep restfully. *Kshama* in Sanskrit means not tying up your minds in knots of peacelessness or resentments. It is like a system of traffic signals and pedestrian crossings and or smooth road turns. It not only saves us from accidents but also guides us to an appropriate path. Mental forgiveness forms foundation for the spiritual and social discipline; it is a grove of serenity where we can meditate, only if purified and free of ambitions. In the absence of forgiveness, it is like the law of fish that prevails: the big fish eats the small fish. Therefore, we must *forget and forgive* and quite simply *live and let live*. Loving kindness is of value only when it has truly taken root in the heart as a permanent condition of being. To share in our friend's joys and sorrows, it is crucial for us in a world agitated by jealousy, resentment and suspicion to find solace. While having lust for own comforts, we are not bothered about knocking on the door of a friend's heart.

Selfless Service or Giving

Ramakrishna had great honor for Sri Ishwarchand Vidyasagar because he was not only an erudite speaker but also had divine grace. Ishwarchand asked how one can achieve "*Sidha*" without "*Sadhna*"? Ramakrishna answered that it was possible by selflessly giving to others. Vivekananda said, "...to be pure and unselfish, and to be good, and doing good to others, is indeed the whole religion". We are happy giving away, rather than getting! We should pray for our enrichment through values like helping, serving, giving etc. to have peace in our life.

Here is a parable about Jesus, who observed two people entering a temple. The rich merchant offered 100 gold coins while the poor old woman put down two mites. Jesus asked his disciples, which of the two made a bigger offering? The obvious answer was the merchant, right? No wrong! It was the woman because the merchant gave what he could spare while the woman offered what she could ill afford to sacrifice.

An episode in Mahabharatha also indicates that giving away possessions acquired through honest means constitutes the highest form of charity. And not the quantity given but with the sincerity, that the charity wins approval of God! Likewise purity of giving lies in the fact when you truly give of yourself. The value of gift is defined by the intention behind it. Krishna analyses (Bhagavad Gita VII 20-22) the act of giving in terms of three gunas-

(i)   From *sattva* "*A gift is (pure) when it is given from the heart to the right person at the right time (of his need) and we expect nothing in return*".

(ii) But when a gift is given in expectation of something in return, or for the sake of a future reward, or when it is given unwillingly, "the gift is of the essence of Rajas, impure He who sees *Shiva* in the poor; in the weak, in the diseased they really worship *Shiva*".

(iii) It is true in Christianity, when Jesus committed them to love and serve by command, "*Love your neighbor as yourself*" for social service. Gandhi, however, gave it a political direction—*giving mutual aid*.

(iv) Krishna also warns that a gift emerges from tamas, or full of it "*When given to a wrong person, at the wrong time and in wrong place, or a gift which does not come from the heart and is given with proud contempt*".

The above points elucidate the ethics of appropriateness and dignity of gift. When it is tainted, it is for hunger of recognition, arrogance, bribe one's way to prestige, it cannot get you grace. In Ancient India, the Hindu ideal of *Seva* or service of human beings in various forms was having maximum religious significance. It was maximally beneficial when moved by compassion in a detached form without expecting any returns. Later it became an element of devotion or *Bhakti in Sikkhism* by Guru Nanak to have bliss. It also was the base of Gandhi's philosophy to serve with the Vedantic idea of oneness of all life, as implicit in Krishna's command in Bhagavad Gita, Worship me in all beings". Ramakrishna had *Seva* or service as the right conduct and service indeed was God-realization. He envisaged a society based on mutual aid or service as sharing by one and all contributing according to his/her ability. Such idea of service is relevant even today particularly when there is no motivation for *Seva*.

All religions must address the different task of restless character of service reaching out to the needy and all others to ensure the unity (in visible form) of all mankind bringing about peace and harmony in the society. Thus according to Mahatma Gandhi, Ahimsa is the attribute of the soul, and should be practiced by everybody in all affairs of human life. If it cannot be practiced thus, it cannot bring peace in society. Gandhi provided the best example of harnessing anger. Like heat, when conserved gets transformed into and transmitted as energy. Likewise anger when controlled can be transformed or transmitted as power to move the world. We can apply this example in our own lives, if we decide to return goodwill for ill will, love for hatred, to create an atmosphere of peace around us through the winning power of nonviolent resistance. This helps to elevate the consciousness of both the oppressed and oppressor.

Peace and Harmony

There are many other ways to deal with conflicts at an individual level, for prop-agating peace and harmony over institutions, among friends, family, societies, nations, in the environment and in the world as a whole.

Peace can be Achieved

To achieve sustainable peace we need to adopt the approaches as below:

a. if we are happy and satisfied
b. if we counter terrorism
c. with self-motivation by changing one self
d. with tolerance or sarva dharma samabhava to all religious
e. it should transcend beyond humanity and should also be achieved with nature
f. with good moral education of human values, like love and respect

g. can be achieved with disarmament
h. with dialogue on solving the problem
i. through the mass media to propagate messages and censor violent contents
j. with Gandhian schemes of nonviolence
k. if we eradicate ignorance, curb our ego, respect all and participate in bringing a change
l. if we work for social welfare and violent groups can be controlled (tamed) by meeting their basic needs. Crime against women shall be reduced
m. by sharing resources and power instead of fighting for and with them.

Resolutions of Peace Summit

N. Radhakrishnan suggests following five points to achieve peace (Peace Summit February, 2015 at Kolkata).

1. Restructure the youth in re-discovering Gandhian campaigns to attract large number of students and youth to the peace initiative;
2. Send "Gandhi Goodwill Missions to SAARC Countries" to promote good relationship;
3. Constitute "Observer Teams" in areas of conflicts and violence;
4. Explore to commence "Shanti Sena Training" in educational institutions;
5. Begin with "Enrolling one million nonviolent Indian homes" as part of inculcating nonviolence in children.

Peaceful living together is possible through harmony and prosperity of individuals or communities or nations. Avoid conflicts and do not resolve them through violence to maintain balance and rhythm.

The following 10 point formula was also resolved at the Peace Summit (Kolkata, 2015) for practicing Non-Violent Principles for both active and passive resistance:

1. In the context of growing violent conflicts in Society, a culture of dialogue for reconciliation has to be promoted along Gandhian lines for furthering peace and harmony.
2. Efforts must be made to promote right understanding of religions and Gandhi s "Sarva- Dharma Sama bhava" that is, equal respect for all Religions, by stressing the common core values of all religions and respecting them.
3. Inter-faith approaches drawing moral, ethical and spiritual values should be promoted for maintaining communal harmony and sustainable peace.
4. Introduce compulsory education for peace and harmony, right from childhood incorporating the ethos and human values of love, compassion, and mutual respect.
5. Adequate and effective steps must be taken to prevent violence against under privileged women and children.

6. Set up structures such as *Shanti-Sena* and the Ministry of Peace in national governments, for promoting social harmony and coherence.
7. Campaigns must be intensified against militarization and military spending by nations.
8. The process of sustainable development and to promote programmes and schemes based on Gandhian Values.
9. Denigration of national leaders like Mahatma Gandhi and distortion of history and the historic role played by national leaders must not be allowed.
10. The great civilizational heritage of India must be stressed in the syllabi of educational institutions.

The world at present is at cross roads because of emerging sociopolitical order, religious and economic chaos. President Obama rightly named it as an era of terror and the menace of war and conflicts. To understand the root cause of such societal unrest, we may be clear that the weapon and vehicle of terror is intolerance and violence, aiming to create fear and destruction that reduces growth and development. This also creates mistrust, and loss of human rights. The legal and sociopolitical reasons that add to violence and terror, and undermine law and order are well-known. It is true that all do not take part in terrorism but simply incitement of terrorism, within a state or country and across the borders, not only kills innocent humans but also impacts law and order and disrupts peace. Globally only a few societies are free from terror; hence a global strategy has to be developed for combating terrorism before it is too late.

Historically we can appreciate that non-violence solutions require spiritual or religious teachings be it adoption and practices of Islam, Jainism, Buddhism, Christianity, Hinduism, Sikkhism and other beliefs. In fact non-violence is innate to humans. It is exemplified best by all non-violent movements that started from time to time, by world leaders, based on spirituality and the religion of love, forgiveness and the loving-kindness for all living beings. These approaches have at their core the basic principle or philosophy of *Ahimsa* or non-violence, shared by Buddhism, Jainism and Hinduism alike. Ahimsa is basically the religion that withstands scientific reasoning of interdependence or the interaction between a web of ecological relationship indicating the effects on one species as being reflected on others. This is rightly elucidated by Lord Krishna in Mahabharta that all worldly affairs are closely related. We, therefore, need to understand that Ahimsa is truly integral to our own co-existence on earth. Dharma is eternal and is not only a law of society or harmony but also the pure reality and the regulatory principle of the Universe.

Non-Violence alone can bring peace by preventing use of physical force or passive acceptance of oppression and also counter the use of arms for struggle or counter it. However, non-violence can bring social change by education, persuasion and or assertion by way of civil disobedience (*Satyagraha*) and the use of media tactics. This has been successfully used in a wide array of problems, viz., Labour agitation, peace keeping, women's movements, and environment protection. Non-violent revolutions have succeeded in 13 nations, including South Africa, Philippines, and of course India, covering almost 60% of the world population.

For this in 1998 the UN proclaimed that the first decade of 21st century (*between 2001–2010*) as the International decade for promoting a 'Culture of Peace and Non-Violence' for the children of the world.

Non violence revolves around three tenets of Gandhi's Satyagraha, (i) *truth,* (ii) *ahimsa (both physical and mental),* (iii) *sacrifice,* that can help individuals with solace and the society to get rejuvenated. Mahatma Gandhi made Satyagraha a potent instrument of immense value to resolve interpersonal social, religious, state or national or even international conflicts. This approach is potent and significantly effective even today to confront terrorism and violence.

The famous quote *"Satyagraha is a weapon of the strong; it admits of non-violence under any circumstances whatever; and it ever insists upon truth"*, clears the doubt with passive resistance as certainly we can overcome violence with clear understanding of love and respect for each other.

Mahatma Gandhi believed that violence will prevail over violence when only someone can prove that darkness can be dispelled by darkness. He also asserted, "We will continue with principle of Ahimsa to resist as has been practiced successfully in many parts of the world". Richard Gregg, a powerful nonviolent crusader, in his book: The Power of Nonviolence (1960), narrates many nonviolent mass movements, e.g. Vaikom Satyagraha in Kerala (India) against caste discrimination that helped the so-called "low caste" communities to obtain the right to walk on roads restricted for them and the right of entry into temples.

Non-violent Resistance Works in Every Situation

People may doubt that non-violence works in every situation and hence are hesitant to strictly adhere to it. Also the unjust apprehension prevails that institutions and others who engage in violent actions easily gain victory. The principle of Nonviolent Resistance can be best articulated as: *To counter and contradict violence boldly with nonviolent initiatives and resolve conflicts by building relationships* based on mutual trust. However, nonviolent principles cultivate passivity among weak societies hence they suffer due to violence, that is difficult to be curbed.

It is proposed that the following few measures can help enhance the power of non-violent principles:

Traditionally the organizers of nonviolent mass movements against social or communal conflicts were always confident of facing the "power" or "ideology" or "system" against which they had to stand. Indeed the nonviolent activists in present day situations must remain alert and active thus not losing their voice, fearing strong opponents. Each conflict needs different, innovative and integrated methods and tools to mobilize the people for successfully implementing nonviolent programmes and to achieve their objectives. Also it is equally important to have an integration of resources, both human and financial, as well as possible collaborations and networking to bring about bilateral cooperation, and dialogues for reaching consensus for the cause of justice and peace.

Conclusion

Peaceful coexistence and cooperation is necessary for harmonious living and prosperity in any society whether for individuals, communities or nations. Conflicts and violence disturb the balance and rhythm of that harmony and create issues against maintenance of peaceful relationships—for human beings to live in peace and to resolve conflicts when threatened. It is therefore the ultimate responsibility of every person to work for a nonkilling and nonviolent world order, and to have peace and justice. Terrorism is a global menace that not only threatens our existence and international peace but also takes away innocent lives indiscriminately, creating fear, hindering societal progress and peaceful existence in liberal democracies.

Our understanding of the present realization supports a view of Ahimsa as self-evident and integral to the existence of life on earth. *Dharma* is eternal, so is *Satya* or Truth, being a fabric that binds human society and therefore the axiom that "anyone who goes against Dharma will only destroy himself." Mahatma Gandhi discovered in the earliest stages of application of Satyagraha, that the pursuit of truth did not admit of violence being inflicted on one's opponent. But he must be freed from error by patience and sympathy. Of course, patience also implies self-suffering! Hence the doctrine came to mean the vindication of truth not by infliction of suffering on the opponent but one's own self.

To conclude I quote Martin Luther King: "Nonviolence means avoiding not only external physical violence but also internal violence of spirit. *You not only refuse to shoot a man but you refuse to hate him.*" Also Mahatma Gandhi's statement is of immense significance even today, when we confront terrorism and violence, when he said, "*Violence will prevail over violence only when someone can prove to me that darkness can be dispelled by darkness.*"

Acknowledgements The author has acquired the Indian philosophy of religion and spirituality for his own life long bringing up in the background of a religious Hindu family and mother having initiated me to the Holy Gita since 1954–55, i.e. early childhood. Besides I've drawn liberally from my readings and learning over the last decades from the TOI volumes of *The Speaking Tree*, and the great Indian guru, Sri Eknath Easwaran, *Living Thoughts of Great People* (Birth Centenary Edition JAICO Publishing House, Mumbai) as well as most importantly from the 2016 book, "Give Nonviolence a Chance: The Journey of Neelkantha Radhakrishnan" by Prof Anoop Swarup, Ed. Konark Publishers Pvt Ltd., NDelhi/Seattle (ISBN 9789322008758).

Part V
Technology

Nanoethics—A Way of Humanization of Technology for the Common Benefit

Štefan Luby and Martina Lubyová

In this paper we summarize a brief history of nanoscience and nanotechnology by documenting the milestones on the roadmap of this branch. We discuss the new properties of materials and structures characteristic for the nanoworld. The role of social sciences and humanities in nanoscience is highlighted. Attention is paid to the new ethical threats originating in the field, as well as to the code of conduct of responsible nanoscientists. These issues represent basic parts of nanoethics—a discipline that opens up a new chapter in ethical studies. In this context we summarize and discuss the present hot topics of nanoethics: human enhancement, health and safety, toxicity of nanostructures and nanotoxicology, violation of privacy, regulations in nanoscience and nanotechnology, specifics of intellectual property rights, technology assessment.

Central to our interests are economic implications, especially the possibility of bridging the nano-divide between the developed and developing world. We point out that nanoscience and nanotechnology open new horizons also in the field of philosophy. There is a need to overcome the problems triggered by nanoscience and nanotechnology reflected in the ambivalence of public opinion towards these new disciplines. This must be provided by correct and open information exchange and continuous technology assessment. Nanoethics does not bring about a new category of ethical problems. Rather, it represents a new manifestation of already known problems and issues. Nanoscience and nanotechnology already conquered the markets. Their main goal should be improvement of the quality of our life.

Š. Luby (✉)
Institute of Physics, Slovak Academy of Sciences, Bratislava, Slovakia
e-mail: stefan.luby@savba.sk

M. Lubyová
Centre for Psychological and Social Sciences, Slovak Academy of Sciences, Bratislava, Slovakia

© Springer International Publishing AG, part of Springer Nature 2019
J. A. S. Kelso (ed.), *Learning To Live Together: Promoting Social Harmony*,
https://doi.org/10.1007/978-3-319-90659-1_21

189

Introduction

Nanotechnology as one cornerstone of NBIC—nano-bio-info-cognitive sciences cluster—is a natural continuation of microtechnology, artificial intelligence and gene technology. The emerging technologies create new challenges in the field of ethics (Sandler 2014). Nanotechnology is a way of control of matter at the nanoscale; nanomaterials and/or nanoparticles are the common examples. New phenomena and qualities in the nanoworld originate from the large surface-to-bulk ratio of nanostructures and small size of nano-entities (Luby 2017). Nanotechnology expands via standard diffusion processes, but also through additional channels, such as meanings, networks, innovations and responsibilities (van Lente et al. 2012). Nanoscience and nanotechnology (N&N) provide numerous opportunities for new materials, medicine and information technology. However, they bring also risks. In order to illustrate the fast progress and main principles of N&N, in Box 1 we summarize selected statements of well-known personalities (Luby 2017; Dosch and van de Voorde 2008; Luby et al. 2015).

> **Box 1 Famous statements about N&N** Only recently we achieved an experimental evidence of the discreetness and granularity of matter, which the atomic hypothesis searched for hundreds and thousands of years.[1]
>
> Wilhelm Ostwald, 1907
>
> The principles of physics do not speak against the possibility of manipulating things atom-by-atom.
>
> Richard Feynman, 1959
>
> Nanotechnology is the art of building devices at the ultimate level of finesse, atom-by-atom.
>
> Richard Smalley, 2000
>
> Biggest breakthroughs in nanotechnology are going to be in the new materials.
>
> Troy Kirkpatrick, General Electric
>
> Nanotechnology in medicine is going to have a major impact on the survival of the human race.
>
> Bernard Marcus, entrepreneur
>
> With inventions we continually underestimate the time span required from idea to their full realization. The new fields of biotechnology and nanotechnology will be evolving for all the years left in the twenty-first century.
>
> John Naisbitt, futurologist

---

[1]In 1905 A. Einstein explained the Brownian movement of small particles on the surface of liquid by their collisions with molecules of the liquid.

Our ability to reap the long-term benefits of nanotechnology will depend on how well industry and governments manage the safety and performance of the first generation of nanotechnology products.

Andrew Maynard, University of Michigan

N&N cover design, preparation and applications of materials and structures having at least one dimension within the interval from 0.1 to 200 nm (1 nm = 10^{-9} m).[2] At the left margin it touches the quantum world. At the right margin the interval adjoins the sub-micrometre region 100–1000 nm (1 μm). Typical nano-objects are molecules, atomic clusters, nano-crystallites, nanoparticles (NP), nanowires, nano-layers, etc.

In Box 2 below we summarize the main milestones in N&N. From about 30 most important milestones published in (Luby 2017) we include only those achieved by Nobel Prize (NbP) and Kavli prize[3] (KP) laureates.

Box 2 The main milestones of N&N

1905 A. Einstein estimated the diameter of sugar molecule at about 1 nm.

1931 M. Knoll and E. Ruska invented electron microscope, NbP for E. Ruska.

1959 R. Feynman presented his talk "Plenty of room at the bottom".

1965 W. Kohn developed the density-functional theory, NbP together with J. A Pople who developed computational methods in quantum chemistry.

1981 G. Binnig and H. Rohrer invented scanning tunnelling microscope, NbP.

1985 R. F. Curl, H. W. Kroto and R. E. Smalley discovered fullerenes, NbP.

1988 A. Fert and P. Grünberg discovered giant magnetoresistance, NbP.

1989 D. M. Eigler arranged Xe atoms on the surface of Ni using the tip of STM. KP with N. C. Seeman for methods to control matter at the nanoscale.

1991 S. Iijima discovered carbon nanotubes, KP together with L. Brus for development of nanoscience of zero and one dimensional nanostructures.

2004 A. Geim and K. Novoselov isolated graphene, NbP.

---

[2]Almost 40 definitions of nanomaterials and nanoparticles were summarized by Boholm and Arvidsson (2016). About 40% of them refer to the nanoscale properties of nano-entities differing from bulk properties, 60% of definitions involve various quantitative boundaries of nanostructure dimensions between 0.1 and 500 nm. Two definitions refer to the surface area of nanostructures >60 m^2/cm^3. According to our opinion the best choice of the upper boundary is 200 nm corresponding to the interface between electron and optical microscope resolution.

[3]The Kavli prize was established by Norwegian entrepreneur and philanthropist Fred Kavli. It is awarded every two years since 2008 in astrophysics, nanoscience and neuroscience.

2014 E. Betzig, S. W. Hell and W. E. Moerner were awarded by NbP for two types of nanoscopes invented between 1989 and 2005. S. V. Hell was awarded by KP together with T. E. Ebbesen and J. B. Pendry for breaking the long-held beliefs about the limitations of the resolution limits of optical microscopy.

2016 J.-P. Sauvage, J. F. Stoddard and B. L. Feringa were awarded by NbP for design and synthesis of molecular machines studied from 1983.

Sources: (Luby 2017; Luby et al. 2015; Stix 2001)

Among the nanotechnology milestones, Feynman's lecture at the Caltech in 1959 is often quoted. Can we really suppose that a vision, although a splendid one, might be a driving force of the progress of research? And, contrarily, can a misleading vision become a decelerating factor? We think that the influence of visions should not be overestimated and the appeal demanding the high responsibility of visionaries (Sand 2016) is exaggerated. In Feynman's case, his personal prestige was a decisive factor that gave importance to his seminal lecture. By all means, the three main events propelling the nanotechnology movement were the achievements of microelectronics, which documented beneficial impact of miniaturization, the inventions of scanning probe microscopes with their abilities to depict and manipulate single atoms, and a new chemistry providing nanomaterials with high reproducibility. Let us quote Feynman in this connection: *Sometimes the truth is discovered first, and the beauty or "necessity" of that truth seen only later* (Feynman 2015).

N&N currently represent a common denominator for the recent developments in chemistry (nanochemistry), biology (nanobiology) and medicine (nanomedicine), whenever these disciplines touch upon the dimension of a single molecule. The interactions obey the laws of physics that are valid up to the level of elementary particles. Therefore we do not speak about nanophysics.

Ethics in science belongs to the common agenda of academics and scholars. During the last ten years nanoethics, i.e. ethics in the rapidly expanding N&N area emerged as a hot topic. In this paper we focus on the most important aspects of nanoethics.

Extrapolation of Microtechnology into the Nanoworld

N&N are a continuation of five decades of miniaturization and the growing level of integration described by Moore's law (Moore 1965). Its formulation was born in 1965 when G. Moore noticed that the number of components per integrated circuit for minimum component cost increased roughly by a factor of 2 annually between 1962 and 1965. He predicted that the rate would remain constant for at least 10 years. His prediction proved to be accurate: the number of transistors in integrated circuits has been doubling approximately every two years up to now. But at present Moore's law approaches its physical barriers. The main barrier will be the heat generation in integrated circuits that will limit their working frequency (Waldrop 2016). The validity of Moore's law is expected to stop around the year 2020 (International Technology Roadmap for Semiconductors 2015). Therefore, the role of N&N grows.

The breakthrough in N&N was the invention of the scanning tunnelling microscope (STM) with atomic resolution (Binnig et al. 1982). In this device the information is not obtained by the use of radiation, but by a mechanical sharp probe scanning over the surface at a very small distance and imaging its topography. Later other types of scanning probe microscopes were created, especially the atomic force microscope. The latter is used also for the manipulation of atoms and molecules. In this sense, Feynman's vision (Feynman 1960) of manipulating matter atom-by-atom has been fulfilled (at least at the laboratory scale, as the process is too slow for practical applications). Thus, the appearance of the STM heralded the emergence of N&N (Moriarty 2001). Nowadays the structures are not synthesized atom-by-atom but from larger blocks, such as nanoparticles composed of thousands of atoms. This additive process is called *bottom–up* approach. It is complementary to the *top–down* formation of structures used in micro- and nanoelectronics.

New Phenomena in the Nanoworld

New qualities of materials and structures that appear in the nanoworld originate from the classical and quantum phenomena. In the realm of classical physics, they are related to:

(a) large surface to bulk ratio;
(b) small size of nano-entities.

Ad (a) In a 1 nm particle almost all atoms are on the surface (Ramachandra Rao et al. 2000). This manifests itself in the decrease of the melting temperature. Gold nanoparticles with a diameter of 2 nm melt already at 500 °C because of the smaller number of bonds to the neighbouring surface atoms. *Per analogiam,* in the nano-world many properties are changing dramatically: insoluble substances become soluble, non-inflammable objects are burning. Chemically toxic elements like As, Cd, Co, show toxicity in bulk. In the nano-world one also speaks about physical toxicity. Large effective surface of structures generates free radicals with threatening consequences, such as inflammations of lungs, fibrosis or tumours (Warheit 2004). It is for this reason that an assertion stating that all nanoparticles are toxic was born.

Ad (b) Small nanoparticles penetrate through membranes, opaque films become transparent, nanoparticle solutions are coloured depending on their size. Another example of utilizing the small dimensions is that of nanoclusters in the matrix of material (e.g. ceramics) that increase its strength (Sajgalik et al. 2006). The most interesting phenomena are introduced through entering the territory of quantum physics. Some of them could be exploited in future quantum computing and quantum encryption (Kaye et al. 2007).

N&N in the Era of Productivity

New technologies usually pass through the cycle of hyperbolic expectations. The five phases of new technology development are (Hersam 2011): launching the technology; over-optimistic expectations; disillusion; enlightenment and realistic views; expectations matched at the level of productivity. Carbon nanotubes can be cited as examples of broadly applied nanomaterial that has been studied already for a quarter of century (Li 2017). The progress of nanomaterials is tremendous in many fields (Dosch and van de Voorde 2008; DG Res. Inov. Ind. Techn 2013): high strength materials, durable composites, surface coatings, improved medicaments, artificial skin, bones, cardiac tissues, plasmonic solar cells, nanosensors, filters for decontamination of water, plant-derived plastic materials (instead of petrochemicals).

Given that currently N&N enters the phase of productivity, the issue of safety and security become the imperative of the day. One has to ask to which extent are N&N harmful. On the one hand, also knives, dynamite or alcohol can be harmful if used in an improper way. On the other hand, there is one significant difference—while a knife is visible to the naked eye, nanoscale entities are invisible to unmediated senses. In this regard (Lingner and Weckert 2016) distant future perspectives of nanotechnologies are confusing because of their realization at nanolevels that are remote from daily experience and individual control.

The public concern is partly focused on the category of sci-fi like self-replication nanosystems, uncontrolled nanorobots or transhumans, but also on real dangers, such as biological warfare, toxic nature of nanomaterials, invasion into privacy—molecularly naked patients, latent fingerprints etc. (Toumey 2007; Ebbesen 2008). In this connection we speak about big brother technologies. (In Sandler (2014) it is stressed that the biggest threat to privacy comes from the government, not from corporations.) Motivated by these concerns, a new Springer journal *Nanoethics* was launched in 2007. A year later the European Commission (EC) adopted a voluntary Code of Conduct (European Commission 2008). The Code is based on a few general principles: research should be comprehensible to the public; research activities should be safe, ethical and contribute to sustainable development; it should anticipate potential environmental, health and safety impacts. (More details are to be found in McGinn (2010).) Short term regulations must be followed by medium and long-term measures reflected in legislation.

Ubiquitous Nanoparticles

N&N work with many types of nanoparticles (Table 1). Some of them are chemically toxic. Threats like cancerogenity, mutagenity, genotoxicity are related to their composition and size. The essential equation reads: *hazard = risk x exposure* (Warheit 2004). The first measure is to avoid exposition via inhalation or skin contact. NPs stored in liquids are less dangerous than aerosols. Safety precautions involve toxicological screening, measuring of particle concentration in the environment, controlling of explosive and flammable mixtures with oxygen and pyrophoric matters (Huber 2005).

On the other hand, nanoparticles can provide a remedy to various diseases or health-related problems. Well-known are the antimicrobial properties of silver NPs. Obviously, the overuse of nanosilver products could enable the flourishing of resistant strains http://www.sciencedaily.com/releases/2012/04/120428000220.htm. A comprehensive study about nanosilver benefits and risks was published by Hansen and Baun (2015): the use of nanosilver has increased dramatically and about 20% of nanoproducts contain nanosilver. It has been exploited in dietary supplements, colours, textile, paper, kitchenware, etc. As the adverse effect of nanosilver in humans bluish-grey discoloration of the skin or eyes, known as agrygia, has been reported after high and repeated oral or respiratory exposure to silver. Unfortunately, the gaps in our knowledge limit the possibilities to assess the potential impact of the increased use of nanosilver on humans. Needless to say, industry and NGOs have conflicting views and interests in this regard.

The Role of Social Sciences and Humanities in N&N

N&N has so far benefited mainly from the results of natural and technical sciences. Nevertheless, social sciences and humanities can increasingly contribute to the development of N&N. The potential fields of intervention include the economic context of N&N, sociology and psychology of consumers, security and protection of privacy–the common denominator of the enumerated activities being ethics (Ebbesen 2008). A qualified overview of ethical and societal implications of nanotechnology has been provided by Allhof et al. (2010). The most frequent issues will be discussed in the following sections.

Table 1 Basic nanoparticles used in N&N (Luby et al. 2015)

Metals	Oxides	Compounds	Semiconductors
Ag, Al, Au, Co, Cu, Ni, Pd, Pt, Zn	Al_2O_3, Fe_2O_3, SiO_2, TiO_2, WO_3, ZnO	AgCu, $BaTiO_3$, CuNi, MoS_2	CdS, CdTe, Si, GaAs, InP

Basic Problems of Ethics in N&N

Human Enhancement (HA)

The enhancement of physical and cognitive abilities of humans by technical and/or biological means challenges ethical, juridicial and religious attitudes. Access to these technologies will be limited due to high cost. They are non-therapeutic and therefore not covered by health insurance (even minor enhancement drugs like Viagra are not covered). Then the standard opinion prevails that HA will impair justice, although in longer horizon enhanced persons should be capable of developing efficient responses to social and other problems (Garcia and Sandler 2008). However, at present also the opposite problem is being discussed, i.e. the dis-enhancement of animals. Emblematic is the blind chicken issue: chicken suffer in high density battery cages where they are prone to cannibalism. The considerations develop even toward the methods aimed at disrupting animals´ ability to experience pain. This approach is repulsive and the dignity of creatures is noted (Thompson 2008). In this regard religious persons worry that HA will contribute to re-defining human nature and diminishing religious understanding (Toumey 2011). Nevertheless, it is doubtful that man can emulate nature, which spent millions of years for perfecting high-performance structures. Moreover, nature operates at room temperature in aqueous ambiance, while chemists operate at high temperatures, in vacuum or in organic solvents (Bensaude-Vincent 2009). Similar reservations could be expressed also towards transhumanism that strives to overcome the human conditions with the aid of new and emerging fields of technoscience including NBIC cluster (Duarte and Park 2014).

Health and Safety

Safety in N&N is related mostly to the fields of health, food and environment. The mapping of early literature in these fields indicates that in the centre of their interest is toxicology (Kjoelberg and Wickson 2007). Nano-entities penetrate the protective mechanisms of the human body, such as the brain-blood barrier and cell membranes. Therefore, they can trigger inflammation and cause damage to DNA (Shrader-Frechette 2007). The cited paper criticizes the insufficient funding for nanotoxicology, as well as issues emanating from conflict of interest: large companies hamper the distribution of information, labelling of products is often incorrect, and data sheets list only restrictions typical for bulk and not for the nano-forms. As *pars pro toto*, the criticism of marketing of TiO_2 nanomaterial in UV protective cosmetics can be mentioned (Jacobs et al. 2010). The process suffered from serious deficiencies, notably the absence of alternatives, low degree of controllability, etc. It seems that this situation will persist because toxicological risks are vaguely defined, difficult to assess, and our knowledge about immune defence response is poor (Roubert et al. 2016). A new journal *NanoImpact* launched by Elsevier since 2016 that focuses on nanosafety research including human nanotoxicology and econanotoxicology is therefore of large importance.

Regulations in N&N

The European Commission has issued a Code of Conduct in N&N. However, its implementation is problematic (del Castillo 2013). The regulation and decision-making done at three levels (European Commission, European Parliament, and Member States) suffers from a lack of effective coordination. The Code is not propagated and applied in the spirit of confidence-building and stakeholders' guidance (Dorbeck-Jung and Shelley-Egan 2013). Although Europe has so far not witnessed a large-scale disaster (comparable to one that China suffered in 2007 Dalton-Brown 2012), this does not mean that people are not intoxicated by N&N products. The opposite was demonstrated at the conference Nanomedicine in 2012 (University of Zürich). Some toxic mechanisms are summarized in Table 2. Nanomedicine is an illustrative example of laws lagging behind increasingly fast-paced scientific technologies (Trisolino 2014).

Intellectual Property Rights (IPR) in N&N

Nanotechnology as a form of molecular manufacturing needs new forms of IPR. The boundary between the invention (which can be patented) and discovery (which cannot be patented) is blurred here (Bastani and Fernandez 2002). Often, N&N involve the sale and distribution of a type or an idea, rather than a token or a device (Koepsell 2009). Last, but not least, the fast inflow of new results suffers from slow patent procedures. In countries where a *grace period* is not implemented (Strauss 2009), patenting opportunities can be lost because of the conflict of interests between publication and patenting. Attention should be paid also to unethical and

Table 2 Toxic mechanisms associated with various nanomaterials (Thomas et al. 2011)

Material	Mechanism
Carbon nanotubes	Inflammation and oxidative DNA damage
Al_2O_3	ROS generation
Au NPs and nanorods	Disruption of protein formation
CdSe	Dissolution of toxic Cd and Se ions
SiO_2	ROS generation, protein unfolding, membrane disruption
CeO_2	Protein aggregation and fibrillation

Note ROS—reactive oxygen species

corrupt practices in scientific publishing, especially to selling and buying papers induced by the publication boom that occurred also in N&N and by promotion requirements (Hvistendahl 2013).

The Economic Context

An overriding concern from the economic point of view is the nano-divide developing between rich and poor worlds. The term nano-divide has been used since 2001. A more politically loaded term is nano-apartheid (Schroeder et al. 2016). As in the case of the so-called digital divide with regard to the information technologies, the bridging of the nano-divide does not seem to be straightforward. This is supported also by the dependence theory in economic development—gains in one region are offset by losses in others.

Different access to nanotechnologies in low-income, middle- and high-income countries is discussed from two basic points (Schroeder et al. 2016):

- the possibility to share nanoproducts;
- the possibility to share nanotechnologies (especially in the field of medicine).

The claims that the affluent world has a moral obligation to help overcome poverty can be found in many development agendas (the Millennium Development Goals, the Poverty Reduction Strategies, the Post 2015 Agenda, and the current Sustainable Development Goals). Nevertheless, one can see that nano-innovations are directed at a high-income world that consumes expensive and luxurious products (such as treatments and cosmetics), while relatively few products are created for the poor parts of the world. The advocates of this divide argue that what initially benefits the rich will be available to the poorer population later on (Trujillo 2015). A more profound analysis of this issue was carried out by Maclurcan (2012). He studied the extent to which nanotechnology facilitates a more equitable world by means of investment policies, international cooperation, frequency of patenting, interviews, etc. He formulated three basic questions and their respective answers:

- Will nano-innovation and innovative capacity be globally and locally decentralised and autonomous?

The findings imply that nanotechnology will continue to concentrate control over innovation, increasing Southern technological dependency and exploitation through the international division of labour.

- Do nanotechnologies offer appropriate technologies for the South?

Although nanotechnology is seen as offering numerous technical advantages, they are oriented away from the Southern needs.

- Do the present and foreseen approaches to nanotechnology governance in the South enable empowering democratic processes?

The governance is focussed on supporting innovations at the expense of public participation and there is evidence of an undemocratic approach.

In summary, nanotechnology is likely to maintain and possibly amplify the inequalities stemming from existing forms of technological innovation. Under these conditions it would be considered a success if inequality and nano-divide does not widen even further (Schroeder et al. 2016). While it is assumed that the Western world will continue to dominate in the field, China and India are likely to narrow the gap. The third-world may challenge the Western culture of individualism by emphasizing compassion (Hongladarom 2009).

Informed Public and Ambivalence of Attitudes

As in the case of many other technologies (such as, for example, nuclear energetics), in N&N one has to deal with large expectations and enthusiastic support on one side, and with concerns about environmental, health and societal effects (leading even to the claims for a moratorium) on the other (Arnaldi and Muratorio 2013). Therefore, openly published information by researchers, journalists, politicians and the public are of tremendous importance. Given that the developing countries consider N&N to be an accelerator of their progress, the vision of researchers is positive, often motivated by the anticipation of new projects and grants. For example, a review of 150 papers published in Brazil shows absolute prevalence of attitudes in favour of N&N (Invernizzi 2008). In the developed world a broad spectrum of attitudes can be found, ranging from enthusiasm to antipathy. This spectrum can be considered as a legacy of the experiences gained in other related contexts, e.g. in the case of biotechnologies (Kearnes and Wynne 2007).

In nanomedicine (Vincent and Loeve 2014) drug targeting is expected to cover 75% of the market. To encourage the public acceptance of N&N, warfare metaphors are used, such as, for example, *therapeutic missiles, nano-bullets, smart bombs*, etc. These are over-simplistic and do not appropriately describe the effective mechanism of the drug.

Technology Assessment

Technology assessment should be done in real time with the aim of shedding light on both its useful and threatening consequences (Guston and Sarewitz 2002). Industrial reports are of considerable importance in this regard, but this information is difficult to access due to high subscription costs (Etheridge et al. 2013). Assessment is particularly important nowadays, when the holistic NBIC concept is being born (Swierstra et al. 2009). New inspiration is provided, for example, by bio-mimetics that studies the properties of living structures and organisms. In this regard, N&N are understood not so much as a starting point of a new industrial revolution, rather as bringing a revolution in the quality of human life (Shelley-Egan 2010).

Conclusions

N&N have already conquered world markets. Therefore, systemization and standardisation of the discipline is necessary. *A sine qua non* condition for achieving further goals is the enhancement of ethical research. From the point of view of ethics, in the N&N field the common belief that science and technology are neutral and only their applications are liable to moral screening is being revised. N&N, however, do not create a new category of ethical problems. Rather, we observe only new manifestation of the problems that are already known (Bacchini 2013).

New qualities in N&N are challenging also from the philosophical point of view. Let us mention complementarity as a basic physical principle: we observe that with the decreasing size of studied structures, the dimensions and complexity of research facilities increase. On the other side the penetration into nano-world facilitates better understanding of the macro-world, even of the Cosmos (Schattenburg 2001).

Symmetry is observed in terms of the dimensions of the current micro- and macro-world research agendas: in the micro-world the manipulation of matter is done at the level of 10^{-9} m, while in the macro-world the operations (e.g. the top achievement of landing on the Moon) took place at the distance of approximately 10^9 m from the Earth. The fact that mankind occupies a central position on this scale underlines our responsibility for fostering the ethical and moral development in N&N and other frontier branches of contemporary research.

Acknowledgements The work was supported by grants: VEGA 2/0010/15, SRDA-14-0891 and CNR-SAS 2016-2018 program.

References

Allhoff, F., Lin, P., Moore, D. (2010). *What is nanotechnology and why does it matter? From science to ethics* (pp. 304). Oxford: Wiley-Blackwell. ISBN 978-1-4051-7545-6.

Arnaldi, S., & Muratorio, A. (2013). Nanotechnology, uncertainty and regulation. *Nanoethics, 7,* 173–175.

Bacchini, F. (2013). Is nanotechnology giving rise to new ethical problems? *Nanoethics, 7,* 107–119.

Bastani, B., & Fernandez, D. (2002). Intellectual property rights in nanotechnology. *Thin Solid Films, 420–421,* 472–477.

Bensaude-Vincent, B. (2009). Self-assembley, self-organization: Nanotechnology and vitalism. *Nanoethics, 3,* 31–42.

Binnig, G., Rohrer, H., Gerber, Ch., & Weibel, E. (1982). Surface studies by scanning tunneling microscope. *Physical Review Letters, 49,* 57–61.

Boholm, M., & Arvidsson, R. (2016). A definition framework for the terms *nanomaterial* and *nanoparticle*. *Nanoethics, 10*, 25–40.

Commission Recommendation on a Code of Conduct for Responsible Nanosciences and Nanotechnology Research, European Commission, Brussels (2008) Document C. 424. http://ec.europa.eu/research/research-eu.

Dalton-Brown, S. (2012). Global ethics and nanotechnology: A comparison of the nanoethics environments in the EU and China. *Nanoethics, 6*, 137–150.

del Castillo, A. M. P. (2013). The European and member states' approaches to the regulating nanomaterials: Two level governance. *Nanoethics, 7*, 189–199.

Dorbeck-Jung, B., & Shelley-Egan, C. (2013). Meta-regulation and nanotechnologies: The challenge of responsibilisation within EC's Code of Conduct for responsible N&N research. *Nanoethics, 7*, 55–68.

Dosch, H., & van de Voorde, M. H. (Eds.). (2008). *Genesys white paper*. Stuttgart: Max-Planck Inst. für Metallforschung. ISBN 978-3-00-027338-4.

Duarte, B. N., & Park, E. (2014). Body, technology and society: A dance of encounters. *Nanoethics, 8*, 259–261.

Ebbesen, M. (2008). The role of the humanities and social sciences in nanotechnology research and development. *Nanoethics, 2*, 1–13.

Eggelson, K. http://www.sciencedaily.com/releases/2012/04/120428000220.htm.

Etheridge, M. L., Campbell, S. A., Erdman, A. G., et al. (2013). The big picture on nanomedicine: The state of investigational and approved nanomedicine products, Nanomedicine: *Biology and Medicine, 9*, 1–14.

Feynman, R. P. (1960). *There's plenty of room at the bottom*. Caltech Engineering and Science J. http://www.zyvex.com/nanotech/feynman.html.

Feynman, M. (Ed.). (2015). *The quotable Feynman* (p. 181). Princeton: Princeton University Press. ISBN 978-0-691-15303-2.

Garcia, T., & Sandler, R. (2008). Enhancing justice? *Nanoethics, 2*, 277–287.

Guston, D. H., & Sarewitz, D. (2002). Real-time technology assessment. *Technology in Society, 24*, 93–109.

Hansen, S. F., & Baun, A. (2015). DPSIR and stakeholder analysis of the use of nanosilver. *Nanoethics, 9*, 297–319.

Hersam, M. (2011). Nanoscience and nanotechnology in the posthype era. *ACS Nano, 5*, 1–2.

Hongladarom, S. (2009). Nanotechnology, development and buddhist values. *Nanoethics, 3*, 97–107.

Huber, D. L. (2005). Synthesis, properties, and applications of iron nanoparticles. *Small, 1*, 482–501.

Hvistendahl, M. (2013). China's publication bazaar. *Science, 342*, 1035–1039.

Invernizzi, N. (2008). Vision of brazilian scientists on nanoscience and nanotechnologies. *Nanoethics, 2*, 133–148.

Jacobs, J. F., van de Poel, I., & Osseweijer, P. (2010). Sunscreens with titanium dioxide (TiO_2) nanoparticles: A societal experiment. *Nanoethics, 4*, 103–113.

Kaye, P., Laflamme, R., & Mosca, M. (2007). *An introduction to quantum computing*. Oxford: Oxford University Press.

Kearnes, M., & Wynne, B. (2007). On nanotechnology and ambivalence: The politics of enthusiasm. *Nanoethics, 1*, 131–142.

Kjoelberg, K., & Wickson, F. (2007). Social and ethical interactions with nano: Mapping the early literature. *Nanoethics, 1*, 89–104.

Koepsell, D. (2009). Let's get smaller: An introduction to transitional issues in nanotech and intellectual property. *Nanoethics, 3*, 157–166.

Li, Y. (2017). The quarter-century anniversary of carbon nanotube research. *ACS Nano, 11*, 1–2.

Lingner, S., & Weckert, J. (2016). Nanoscale-technologies as subject of responsible research and innovation. *Nanoethics, 10*, 173–176.

Luby, S. (2017). *Nanoworld at the palm of your hand, VEDA*. Bratislava: Publ. House of SAS. in press.

Luby, Š., Lubyová, M., Šiffalovič, P., Jergel, M., Majková, E. (2015). A brief history of nanoscience and foresight in nanotechnology, Chap. 4. In M. Bardosova, T. Wagner (Eds.), *Nanomaterials and Nanoarchitectures. Proceedings of NATO ASI on Nanomaterials and Nanoarchitectures, Cork, July 2013* (pp. 63–85). Dordrecht: Springer. ISBN 978-94-017-9937-9.

Maclurcan, D. (2012). *Nanotechnology and global equality* (pp. 451), Singapore: Pan Stanford Publ. ISBN 978-981-4303-39-2.

McGinn, R. (2010). Ethical responsibilities of nanotechnology researchers: A short guide. *Nanoethics, 4*, 1–12.

Moore, G. E. (1965). Cramming more components onto integrated circuit. In *Proceedings of Electronics*, April 19 (p. 114).

Moriarty, P. (2001). Nanostructured materials. *Reports on Progress in Physics, 64*, 297–381.

Nanotechnology: The invisible giant tackling Europe's future challenges (2013). DG Res. Inov. Ind. Techn. EUR 13325 EN. ISBN 978-92-79-28892-0.

Ramachandra Rao, C. N., Kulkarni, G. U., Thomas, P. J., et al. (2000). Metal nanoparticles and their assemblies. *Chemical Society Reviews, 29*, 27–35.

Roubert, F., Beuzelin-Ollivier, M.-G., Hofmann-Amtenbrink, M., Hofman, H., & Hool, A. (2016). Nanostandardization in action: Implementing standardization process in a multidisciplinary nanoparticle-based research and development project. *Nanoethics, 10*, 41–62.

Sajgalik, P., Hnatko, M., Lences, Z., et al. (2006). In situ preparation of Si_3N_4/SiC nanocomposites for cutting tools. *International Journal of Applied Ceramic Technology, 3*, 41–46.

Sand, M. (2016). Resonsibility of visioneering—opening Pandora's box. *Nanoethics, 10*, 75–86.

Sandler, R. L. (Ed.). (2014). *Ethics and emerging technologies* (pp. 583). New York: Palgrave Macmillan. ISBN 978-0-230-36702-9.

Schattenburg, M. L. (2001). From nanometers to gigaparsecs: The role of nanostructures in unraveling the mysteries of the cosmos. *Journal of Vacuum Science and Technology B, 19*, 2319–2328.

Schroeder, D., Dalton-Brown, S., Schrempf, B., & Kaplan, D. (2016). Responsible, inclusive innovation and the nano-divide. *Nanoethics, 10*, 177–188.

Shelley-Egan, C. (2010). The ambivalence of promising technology. *Nanoethics, 4*, 183–189.

Shrader-Frechette, K. (2007). Nanotoxicology and ethical conditions for informed consent. *Nanoethics, 1*, 47–56.

Stix, G. (2001). *Little big science*. Scientific American, 26 September 2001.

Strauss, J. (2009). Is the patent system fit to meet the needs of the "triple helix" alliance?. In M. Eder (Eds.), *20 Jahre Europäische Akademie der Wissenschaften und Künste* (Weimar ed., pp. 371–385) Salzburg. ISBN 978-3-89739-666-1.

Swierstra, A. T., Boenink, M., Walhout, B., & van Est, R. (2009). Converging technologies, shifting boundaries. *Nanoethics, 3*, 213–216.

Thomas, C. R., et al. (2011). Nanomaterials in the environment: From materials to high-throughput screening to organisms. *ACS Nano, 5*, 13–20.

Thompson, P. B. (2008). The opposite of human enhancement: Nanotechnology and the blind chicken problem. *Nanoethics, 2*, 305–316.

Toumey, C. (2007). Privacy in the shadow of nanotechnology. *Nanoethics, 1*, 211–222.

Toumey, C. (2011). Seven religious reactions to nanotechnology. *Nanoethics, 5*, 251–267.

Transistors won't shrink beyond 2021 (2015). International Technology Roadmap for Semiconductors.

Trisolino, A. (2014). Nanomedicine: Building a bridge between science nad law. *Nanoethics*. https://doi.org/10.1007/s11569-014-0193-y.

Trujillo, L. Y. C. (2015). In: R. L. Sandler (Ed.), *Ethics and emerging technologies* (Vol. 9, pp. 251–254). Nanoethics, 2014. New York: Palgrave Macmillan.

van Lente, H., Coenen, C., Fleischer, T., et al. (Eds.). (2012). *Little by little: Expansion of nanoscience and emerging technologies* (pp. 225). GmbH, Heidelberg: Akad. Verlagsges. ISBN 978-89838-674-6.

Vincent, B. B., & Loeve, S. (2014). Metaphors in nanomedicine: The case of targeted drug delivery. *Nanoethics, 8*, 1–17.

Waldrop, M. (2016). More than Moore. *Nature, 530*(11.2.), 144–147.

Warheit, D. B. (2004). Nanoparticles: Health impact? *Materials Today, 7*(2), 32–35.

Science, Technology and Innovation (STI) in the Age of Globalization

Aderemi Kuku

Globalization and Consequences for Mathematics, Science, Technology and Innovation Development

Globalization is the process by which nations and peoples become increasingly interdependent resulting in increased international flow of knowledge, technology, political and socio-economic ideas, human capital, etc. As such, it has promoted unity in diversity and turned the world into a global village.

Some Advantages of Globalization Include the Following

i. Globalization promotes free international trade whereby some countries don't impose taxes, duties, or quotas on the import of goods from other countries thus enabling consumers to buy goods at relatively low cost.
ii. It improves global connectivity of the internet.
iii. It promotes employment opportunities in developing countries due to the emergence of new companies and new markets thus helping to improve the standard of living and alleviating poverty.

Some Disadvantages of Globalization

(i) Globalization has increased Carbon Dioxide emission through increased world coal burning etc. For example, since China joined the World Trade organization in 2001, its coal use has increased phenomenally.

A. Kuku (✉)
African Academy of Sciences, National Mathematical Centre, Abuja, Nigeria
e-mail: aderemikuku@yahoo.com

© Springer International Publishing AG, part of Springer Nature 2019
J. A. S. Kelso (ed.), *Learning To Live Together: Promoting Social Harmony*,
https://doi.org/10.1007/978-3-319-90659-1_22

(ii) It has caused inflation in many countries because of increased demand for food and energy resulting in phenomenal rise in commodity prizes.

(iii) It has caused un-employment in developed countries because it has transferred jobs to developing countries where labor is cheaper through out-sourcing of jobs.

Consequences of Globalization for STI Development

Global challenges require global focus for solutions and it has become increasingly possible to generate global ways of attacking the challenges in this age of globalization than in the past. Most of the challenges in the age of globalization can be translated into challenges involving science and technology which invariably involve challenges in the mathematical sciences since mathematics is the bedrock of all development in science and technology. For example, many areas of science and technology which need mathematical sciences for their in depth study also pose global challenges that need to be globally tackled and solved across geopolitical, linguistic and cultural boundaries (see section in "Some Impact of Mathematics on Other Areas of Science and Technology"). As such, the age of globalization is facilitating finding solutions to such problems.

Nature of Mathematical Sciences Vis-à-Vis Other Areas of Science and Technology

Nature of Mathematics

Mathematics is a Heritage of All Mankind

It has a very long history running into thousands of years with contributions from various cultures-Sub-Saharan Africa, Egyptians, Babylonians, Greeks, Arab-Islamic, Indians, Mayans, Chinese, Arabs, Europeans, etc. However, the last 600 years have been dominated by Europeans and cultural associates like USA, Australia, Canada, etc. refining the subject into a form suitable for Science, Technology and Socio-Economic Development.

Various Ramifications of mathematics include the following;

- NUMBER—which involves counting; measurements (e.g. of length, weights); understanding of integers; rational, real, complex, p-adic numbers, etc.

- SHAPE—which leads to studies in geometries, topology, Lie groups with applications, gauge field theories, fractals, catastrophes, attractors, etc.

- MOVEMENT—of waves, planets involving ODE, PDE, Fourier analysis, calculus of variations.

- CHANCE AND RANDOMNESS—with associated mathematics e.g. probability, statistics, stochastic differential equations, etc.-all with added exploratory and processing powers of new technologies like computers.

Contemporary Methods in Mathematics Are Rather Profound, Sophisticated, Technical & Diversified.

For example, an easily stated problem like 'Fermat's last theorem' has so far been solvable by highly sophisticated and abstract techniques. Consequently, we have global illiteracy in contemporary mathematics resulting in hostility from various Government Institutions, private sectors and the public. We thus have serious pedagogical issues about teaching and learning contemporary materials in mathematics.

The Four Major Components of Science and Technology

Basic Sciences which consist of;

Mathematics (including statistics and computer science); Physics; Chemistry; Biology (including basic medical sciences)

Applied Sciences which include:

Medicine and health; Agriculture (including livestock, fisheries and forestry studies)

Earth Sciences (including Meteorology, Oceanography, irrigation and soils, minerals exploration, etc.)

Lower or Classical Technologies including:

Iron, steel and other metal goods; Petroleum technologies; Power generation and transmission as well as Design and fabrication industries

High Technologies which include;

Micro-electronics (including software development, fabrication of microchips with industrial application, computer-aided design, etc.); Space Sciences; New Materials (including high temperature super-conductors...), Pharmaceuticals and fine chemicals; Biotechnology (including molecular biology, genetics and microbiology-useful in agriculture, energy, medicine)

Science and Technology as concentric layers

Hence, Science and Technology could be viewed as concentric layers with DIFFUSE BOUNDARIES with a central core of basic sciences and mathematics at its inner most core such that theories from the inner core help to solve problems in applied sciences as well as technology while problems arising from outer layers of technology and applied sciences provide the inner cores of basic sciences and mathematics with new structures, new concepts and new methods.

Some Impact of Mathematics on Other Areas of Science and Technology

1. **Electrical Generation Technology**
   Inspired by Faraday's Theory of electricity and magnetism
2. **Wave Propagation, x-rays, radios, television, oil exploration, etc.**
   Makes use of Maxwell Equations, Fourier Analysis..
3. **Acoustics, electric currents in the brain, turbulence, stellar structures, etc.**
   Fourier Analysis, Wavelet Analysis, etc.
4. **Computer Revolution is Mathematics Revolution.**

    i. **Computers are a creation of Mathematics**–Alan Turing's cogent and complete analysis of the notion of computation and logical proof of the existence of the universal computer.
    ii. **Computers are also creators of new areas of mathematics**: for example, complexity theory, proof theory and theory of Algorithms.
    iii. **Computers have recorded tremendous success in the solution of outstanding mathematical problems** e.g. four color problem; classification of simple groups.
    iv. **Computers are useful for teaching mathematics**–calculus, matrix algebra, probability, statistics, geometry etc. Computers are also useful in solving problems arising in technology—commerce, business, economics. Computerization of essential services—payment of salaries, banking and library services, etc., have made life easier.

5. **Subatomic Particles, Crystallography, Photochemistry** etc. Group theory
6. **Tracing of Hurricanes, Studying Aircraft Flight Shocks in Non-Linear Waves**, etc. Partial Differential Equations
7. **Communication, Urban Planning, Neurophysiology** Graph theory, Network science
8. **Computational Models of the Heart, Kidney, Pancreas, Ear** P.D.E's, Physiological fluid dynamics.
9. **Green House Effec**t Numerical Solution of PDE's
10. **Trajectories of Celestial Bodies, Meteorology**—Dynamical Systems. ODE, PDE, Hamiltonian Mechanics
11. **Population Biology, Genetic Engineering**—Probability, Dynamical Systems, Wave Propagation
12. **The Role of ICT and Artificial Intelligence.**

The study of all areas mentioned so far, apart from requiring Mathematics, also have strong ICT components. The ICT components of various areas of science and technology have become more pronounced and prominent in this age of globalization. Indeed, most recent developments all over the world have been ICT driven. Moreover, ICT has been used effectively to gather, manipulate, present, and communicate information all over the world.

More specifically, ICT has been particularly useful for:

- Bridging the technological divide between developed and developing countries.
- Significant growth in the internet access all over the world,
- Unifying themes in S & T and in various disciplines, which must affect curriculum at all levels. e.g. education, health, agriculture, economics, and commerce. Most global challenges can be translated into challenges involving S & T, and ICT, hence mathematical sciences.

Robotic Technologies are based on Artificial Intelligence technology with human-like capabilities to learn, solve problems, and also be creative. These technologies are used for many purposes—transportation, industrial, healthcare, education, agriculture, concept of autonomous vehicles, surgery, therapeutic care, elderly care, etc. As more and more jobs of the future are done by robots, it will be inevitable that the future offers jobs mainly to those who are ready to absorb and implement new ideas and solve unconventional problems.

Current and Future Prospects of STI

We now discuss the current and future prospects for some areas of science and technology that have been prominent and topical:

1. Health and Wellbeing
2. Climate Change
3. Sustainable Agriculture and Food Security
4. Renewable Energy
5. Water and Sanitation

Health and Well Being—the World Is Full of Diseases to Be Cured or Prevented

Children and adults in the world are being infected by diseases of various types especially in developing countries which have suffered or are still suffering from such diseases as malaria, cancer, measles, TB, Ebola, Polio, Yellow Fever, Tetanus,

HIV, Malnutrition. Human health continues to be endangered due to cross infections from animals and plants creating challenges that have to be tackled globally.

Through international cooperation from all concerned e.g. WHO, Universities, Research Institutes, Academies, Professional Associations, Pharmaceutical Industries and various Foundations etc. some of the diseases are being eradicated or are close to being eradicated.

Can we ever envisage a world free of diseases? Certainly not.

However, Mark Zukerman has an ambitious project and has invested billions of dollars to eradicate all diseases by the end of this century through getting the best minds to create basic scientific research tools, possibly hardware and software.

Problems connected with solving health problems in developing countries include the following:

- Health is grossly underfunded in terms of prevention, health care delivery and research.
- Poverty and malnutrition result in poor health
- Health insurance is not available to most citizens

The future is difficult to predict since new diseases keep surfacing from time to time. Suffice it to say that the world is alert and will continue to be alert to unexpected outbreaks. The developing countries of today still have a long way to go to invest heavily on the health of their citizens.

Climate Change

Climate change has many negative consequences—stress leading to cardiovascular and respiratory diseases, inadequate food availability, extreme weather conditions resulting in floods and drought. High temperatures linked to climate change are identified with frequent fires that have destroyed life and property and resulting in air pollution. Warmer temperatures also cause rising sea level and increased floods and droughts that negatively affect the quality and availability of clean water as well as sanitation systems.

Examples: Climate change has threatened populations and property around Mount Elgon on Kenya-Uganda border. Moreover, high population densities, unsustainable land use practices and heavy rainfall make the mountain more prone to landslides and downstream flooding that destroy life and property.

Active ice melt could trigger uncontrollable climate change-"tipping points"—in the active region that could have catastrophic effect around the globe.

NOTE: climate "tipping points" occur when a natural system such as polar ice cap undergoes a sudden overwhelming change that has a profound effect on surrounding ecosystems.

In Africa, air pollution from various sources—burning of rubbish, cooking indoors with inefficient fuel stoves, millions of electricity power generators for homes and industries have been rather deadly—releasing to the atmosphere dangerous gases such as sulphur dioxide, carbon monoxide etc. Indeed, from 1990–2013, air pollution increased by 36% and death from air pollution increased by 18% according to OECD—Organisation for Economic Cooperation and Development)

Sustainable Agriculture and Food Security

The last 100 years have witnessed the evolution of agriculture into highly-mechanized farming, with various types of storage facilities and scientific improvement of seeds and crops. Yet most of the teeming world populations are hungry.

Prospects for the future include current and future scientific research efforts on robotic farmers, self-driving tractors, further new scientific blending and enrichment of seeds and crops. Are we sure of availability of food for the estimated ten billion inhabitants of our planet by the end of this century?

Renewable Energies

It is well known that fossil fuels are also responsible for a lot of land, water, air pollution beyond the CO2 emissions, and are largely responsible for climate change. Hence the current efforts to invest in renewable energy resources e.g. solar, wind, that are clean and will not run out, and are low in carbon energy resources. So, there are lots of studies going on concerning the current and future supply of electricity through effective and low carbon lithium batteries.

Example: Morocco, which has no fossil fuel reserves and relies on imports, wants to increase its renewable electricity generation capacity 52% by 2030. So, Morocco is planning in the next few years to switch on the first phase of the world's largest solar plant that will provide electricity for 1.1 million people

Water and Sanitation

Scarcity of potent water is already topical as a problem requiring global attention and there are various research efforts to ensure availability of clean and potent water the world over.

Warmer temperatures, rising sea levels, increasing floods, droughts and melting ice affect the quality and availability of water as well as sanitation systems.

Population growth, increase in water consumption, higher demand for water supply due to industrialization and urbanization are draining water resources worldwide.

All these factors force people especially in developing countries to use unsafe water which expose them to potentially deadly diseases like cholera and diarrhea.

Nearly 600 million children (one in four children worldwide) will live in areas with limited water resources by 2040 (UNICEF). Climate change will intensify these risks of depletion of sources of safe water in the coming years.

In sub—Saharan Africa, 70% of the population has no access to potent drinking water. The livelihood of more than 30 million people depend on Lake Victoria but water pollution, resource exploitation and the region's dramatic population growth are putting extreme pressure on the lake's ecosystem. Countries around the lake have to work together to contain the situation.

In Latina America, one of the world's water-richest regions, distribution is rather uneven. This again is a situation requiring co-operation of neighboring countries for equitable distribution.

In South East Asia, the problem is acute. However, China's power display creates a lot of conflicts between countries. This is because China builds a big dam across the Makong river and can turn off the tap for countries downstream. So, all countries in the neighborhood of the Makong river have to work together.

The Way Forward

All Governments, especially in developing countries should:

1. Demonstrate more Political Will for Radical Increase in Funding for Science, Technology and Innovation
2. Inculcate Scientific Culture in the citizenry through popularization of Science.
3. Effect Radical Improvements in Teaching and Research Infrastructures and Facilities.
4. Enhance closer Links Between Universities, Research Institutes and Industries.
5. Attract Good Students and Personnel for Careers in Mathematics, Science and Technology.
6. Strengthen Existing Centers of Excellence and Create New Ones.
7. Stem Brain Drain in Mathematics, Science and Technology and turn Brain Drain into Brain Gain.
8. Improve radically the Level of Scientific Research Outputs.
9. Leapfrog into Scientific Frontiers and High Technologies.
10. Promote and develop capacity building and a critical mass of specialists in all areas of Science and Technology.

Cryptocurrencies: Can They Live Together with National Currencies and What Impact Do They Have on National and Global Economies?

Konny Light

The evolution of societies and technology has transformed our medium of exchange from shells to metal to plastic, and now to cryptocurrencies. Cryptocurrencies have grown by nearly 40 times in the last two years, reaching a capitalization of $675 billion and causing both national and international impact in financial markets. Their faster, less costly, less regulated, and infinitely less fraud-prone management is causing their underlying blockchain algorithms to transform financial markets and potentially banking and financial regulations worldwide. Many technical, legal, economic, and social issues have been raised from their popularization and the fact that—by their nature—they are international. While an overview of the cryptocurrency landscape and their trends are discussed, the reader is warned that this financial niche is transitioning so fast that some facts herein may be outdated by the time of its publication. Recommendations for the healthy use and regulation of cryptocurrencies are set forth with suggestions of the larger underlying issues to be resolved.

K. Light, JD (✉)
International S.T.E.P.S. Foundation, Chania, Greece
e-mail: KLight777@msn.com

© Springer International Publishing AG, part of Springer Nature 2019
J. A. S. Kelso (ed.), *Learning To Live Together: Promoting Social Harmony*,
https://doi.org/10.1007/978-3-319-90659-1_23

Introduction

Currencies can generally be defined as a medium of exchange with extrinsic, or implied, value that is not necessarily determined by the physical characteristics or utility of the medium itself (Gross 2014). While the form of currency may have changed from shells to metals to plastic, the underlying purposes that fuel our monetary system remain constant. The Director of Research for the International Monetary Fund reported this year that standard models of macroeconomic stabilization that failed the world in the 2008–9 banking crisis need to be adjusted to incorporate "risky" activities such as non-regulated cryptocurrencies in order for these models to be a meaningful tool in the management of economies and the promulgation of effective policies (Obstfeld and Taylor 2017. See also, 1998 and 2007).

Unlike nationalized currencies like the dollar, pound sterling, and yuan, cryptocurrencies are "mined" or processed at a mathematically-controlled rate through decentralized, random computers, subject to free market demand and prescribed ceilings. This distinguishes them from traditional currencies, which follow the decision-making of central banks. The cryptography inherent in cryptocurrency also makes it more anonymous than any real or virtual currencies, which are tracked by banks and developers, respectively (Wagner 2014).

Additional advantages of "cryptos" for commercial transactions over national currencies, include:

- Primarily operate outside third party financial institutions,
- Transactions are processed without regard to national borders,
- Unregulated in most jurisdictions until very recently,
- Hence no government or third-party reporting, even on large transactions,
- The transactions can be processed in minutes rather than 3–10 bank days,
- Users on the system are identifiable only by their virtual addresses, and
- Cryptocurrencies for the most part can only be manipulated by its stakeholders, not by lobbyists or politicians (Obstfeld and Taylor 2017).

Since the traditional financial system has been slow to recognize cryptocurrencies and to adopt its underlying technical innovation, a unique nomenclature has arisen to describe and name the facets of cryptocurrency dealings. For example, since banks do not take deposits in cryptocurrencies, they are stored in virtual "wallets" that inventory and track one's various cryptocurrencies and related transactions. These wallets are software programs that can be loaded onto a computer or smartphone. Some wallets have "exchanges" that can actually exchange the cryptocurrencies into other national/fiat or crypto-currencies. Unlike wallets, exchanges do typically have encryption key codes for each transaction and agreements that feed directly into the traditional banking or financial system.

Unlike traditional currency, cryptocurrencies can be earned by participating (i.e.; by lending one's computing power to be used) in the underlying "blockchain" that is created to randomly spread multiple validations of diversified ledger components of transactions. The innovations of cryptocurrencies raise significant theoretical, legal, and practical issues that may well make box store banks extinct, reduce the role of central banks, and favor world bank expansion.

History of Cryptocurrencies: Currencies of the Future?

Most articles credit a Japanese mathematician with the fictional name of Satoshi Nakamoto with the creation of the first cryptocurrency. However, Nakamoto actually developed an algorithm for diversified ledgers. Diversified ledgers had been used for years by the insurance and other industries to facilitate the intricate details of shifting financial risk under differing conditions and times in the process of a transaction, such as when an insurance binder is issued, when it is signed by the customer, when the customer's payment is processed, when the policy is issued, and when a claim occurs. When Nakamoto's algorithm was paired with Ron Rivest's public key cryptography (1977), David Chaum's anonymous internet payments (1993), and Hal Finney's proof-of-work token money (2004), the world had a solution to the global banking crisis of 2008–9 in which the fraud and mismanagement of a central banking entity and the lack of oversight by the government could not collapse the world economy (GitHub 2017). This convergence of progressive innovations and global crises led to the launch of the Bitcoin[TM]—the first cryptocurrency.

The cryptocurrency system uses a network of randomly connected computers that verify and validate online transactions with digital ledgers. The community of online computer participants are known as "miners" and they earn cryptocurrencies by their mining (Gross 2014). While the majority of miners are in Asia, they are distributed all over the world (Hileman and Rauchs 2017).

As of April 2017, the following cryptocurrencies are the largest after Bitcoin in terms of market capitalization: Ethereum, Dash, Monero, Ripple, and Litecoin. Ethereum (ETH) features a decentralized blockchain platform with its own Turing-complete programming language whereby contracts/records are executed by every node. Officially launched in 2015, Ethereum has attracted significant interest from many developers and financial institutions. Dash, a privacy-focused cryptocurrency, launched in early 2014 with block rewards being equally shared between miners and 'masternodes'. Monero (XMR) cryptocurrency system is another 2014 privacy oriented currency focused on hiding the origin, transaction amount, and destination of transacted coins. Launched in 2014, Ripple (XRP), a non-blockchain cryptocurrency, supports a 'global consensus ledger' that is utilized by financial institutions and money service businesses. Litecoin (LTC) was launched in 2011 and is similar to Bitcoin, with an altered mining algorithm (McKinsey & Co. 2017).

Trends in Cryptocurrencies

Global accounting firms advise that banks may go the way of bookstores (i.e.; fold their operations) because they are too cumbersome and too costly in processing simple financial transactions in light of the decentralized processing of cryptocurrencies. (KPMG 2015). Innovation in money services, such as Square, cryptocurrency exchanges and smartphone money exchange apps are appealing to the young consumer over box store banks.

The combined market capitalization (i.e., market price multiplied by the number of existing currency units) of all cryptocurrencies increased more than threefold from early 2016 to April 2017 and skyrocketed by 38-fold by mid-December 2017 with a market capitalization of nearly three quarters of a trillion dollars (MoneyControl.com 2017). With the price of Bitcoin having gone from approximately $680 in early 2016 to over $20,000 a coin as of December 31, 2017, the market is showing a confidence and respect that only a few governments are mirroring. This is a sea change in light of the fact that five major banks in Europe did only $1.6 billion a quarter, or approximately $6.4 billion combined, in card transactions in a recent year (KPMG 2017).

What is equally significant is that Bitcoin's share of the cryptocurrency market capitalization has declined by more than 60% as new and improved cryptocurrencies emerged (Hileman et al. 2017). A survey of 2,000 British consumers by Kalixa Pro in June 2014 indicates that they are becoming a cashless society (KPMG 2017). Cryptocurrencies are expanding their outlets of exchange as large retailers such as Dell, Expedia, and Overstock accept Bitcoins.

"Bitcoin is the internet of money" according to Jon Malonis, the Director of the Bitcoin Foundation. He predicts that banks will step into the e-wallet and exchange businesses along with taking over merchant processing and escrow services for bitcoin transactions. Once perceived as a threat to the traditional banking industry, a plethora (12,000 in all) of FinTech start-ups are being sought to partner with established banks and financial stalwarts (KPMG 2017).

Similarly, since the U.S. unpegged the dollar from the gold standard, the consumer relationship with cash is more virtualized, leaving people open to earlier adoption of new financial technologies. This may be one reason why, when Price, Waterhouse, Cooper (PwC), one of the top 5 global accounting and financial consulting firms surveyed cryptocurrency users, it discovered that 86% expect their use of it to significantly increase in the next three years. The most popular use (81% of survey respondents) is "online shopping" (PwC 2015).

The industry trend is to balance emerging technologies with traditional financial services. More importantly, blockchain technology has enabled many emerging economies to completely skip entire stages of development. For example, just as cell phones made it unnecessary for African countries to build telephone lines, Senegal and Tunisia are creating their own cryptocurrencies, potentially eliminating the need to build financial infrastructures, clearing houses, and other intermediaries (Heathman 2017). The transnational nature of the diversified ledgering behind cryptocurrencies may well indicate that we are at the beginnings of a world currency, as trusted international networks are being established outside of the political arena to validate national and international financial transactions.

Regulation of Cryptocurrencies

Though it is not clear what qualifies as legal tender, "[t]he history of regulating money and legal tender suggests that it is not likely that governments will surrender their privileges to regulate cryptocurrency issuers, exchanges, administrators, or users" (Gross 2014). According to *BitLegal* 54 nations allow the use of bitcoins, while Vietnam and Iceland are "hostile" to them (Yap 2016).

Asian regulators have been the most active. China periodically cracks down on the use of bitcoin, sending the currency valuation plummeting, later relaxes and the Bitcoin rises—and perhaps for that very reason—it has recently decided to issue its own cryptocurrency. Korea relaxed its capital requirements for cryptocurrency-related companies, and the investor rush was so frenzied that they reversed their regulation and recently banned trading. However, Korea, too, is said to be considering a crypto or alternative currency (Hendrickson et al. 2016).

Japan's new virtual currency legislation takes an old line approach requiring registration with the Financial Services Agency (JFSA), but only allows firms that meet the JFSA standards of sufficient staff and capital. This is Japan's way to address digital currency fraud (i.e.; where individuals have been sold coins with no value by some providers), anti-money laundering (AML), and know-your-customer (KYC) fraud prevention. Under the new regime, cryptocurrency exchangers no longer have to pay an 8% tax, but must explain the workings of Bitcoin trading to customers. Even though it has shaken the very foundation of cryptocurrencies—their anonymity—Japan requires adherence to strict AML/KYC rules requiring that exchangers post hard copy documents to customers for even online transactions (Casey 2014).

The regulatory environment is the largest factor in the growth (with no regulation) and volatility (with "banning" regulation) of cryptocurrencies. However, regulations to date have focused on currency services rather than on the miners who make up the blockchain. The largest thefts to date of cryptocurrency has been the hacking of Mt. Gox in 2011 and Coincheck in 2018. Still the cryptocurrency ecosystem is open to manipulation by miners seeking to override other miners' blockchains (Norry 2017). Selfish miners (those overriding or pooling to beat out other blockchains) are a real threat to the very security that cryptocurrencies have brought to the financial processing arena, yet regulators have not even begun to address these greater threats (Böhme 2015). Other issues to be resolved by laws and regulators are decentralized ownership, international jurisdiction, user anonymity, and blockchains of nondigitized assets requiring legal consideration of off-chain settlement (Hendrickson et al. 2016).

Emerging Issues in Cryptocurrencies

Only now that cryptocurrency's capitalization has reached the heights it has are countries realizing that they have missed out on a tremendous tax base. A major issue then becomes how the government will value the coins and their underlying transactions.

Other fundamental issues exist around the right of citizens to contract and at what point the government is justified in regulating those contracts. A cryptocurrency's algorithm cannot be modified by a central monetary authority—government-backed or otherwise—without the consent of a majority of users on the system. So to the extent that national currencies are replaced with bitcoin and other cryptocurrencies, the relevant monetary authority loses control of the total money supply and thus may not accomplish its monetary policy goals. While the lack of monetary control was considered insignificant when the market capitalization for cryptocurrencies was $5.4 billion in 2015, the current market cap of $675 billion is expected to impact every major economy. Similarly, to the extent that investment is put into cryptocurrencies as opposed to government bonds and assets, it would reduce the government's ability to raise revenues from seigniorage under the current lack of regulation in most jurisdictions.

Another major issue is the anonymous transaction of illegal business via the use of cryptocurrencies. For example, many gambling, weapon, and questionable goods/services websites now accept cryptocurrencies. It was the public crackdown on illegal and illicit websites and businesses that accepted cryptocurrencies that initially gave the alternative currencies a bad name (Hendrickson et al. 2016). However, the cryptocurrency market is currently growing at such an exponential rate that the financial system is seeking a way to legitimize it despite the potential for illegal use. The question then becomes, how do we keep the benefits of cryptocurrencies' lower costs, greater accuracy, and immutable financial data, and still allow enough regulation to prohibit or deter illegal and illicit uses? (Han et al. 2016) Since the cryptocurrency platform easily allows for cross-border transfer of wealth and services, is it not time for a new global monetary fund that mirrors the stakeholder input model achieved by the evolution of cryptocurrencies (i.e.; each cryptocurrency can be changed only by input of all stakeholders)?

While there is great geographic dispersion of cryptocurrency miners around the world, there exists a large concentration in China. Should the world financial system start to employ anti-trust laws to thwart the dangers of monopolization since a key component of the reliability and non-hackability of the cryptocurrencies is the diversity, honesty, and randomness of its miners?

Potential Impact on National Economies

Many governments are concerned that crypto-currencies will be used for illegal transactions and will hinder the management and control of money within their country, as well as the ability to track and collect tax revenues. Central bank representatives have stated that the adoption of cryptocurrencies, such as Bitcoin, poses a significant challenge to central banks' ability to influence the price of credit for the whole economy. Their fear is that as the use of cryptocurrencies increases, there is bound to be a loss of consumer confidence in fiat (government) currencies (Hendrickson et al. 2016). Perhaps this is a reason for governments to give more "official" recognition to cryptocurrencies and value them in relation to their fiat currencies. Cryptocurrencies are a more true reflection of what the populace values as exchange and are both more up-to-date and volatile.

While some have opined that widespread use of cryptocurrency would make it more difficult to gather data on economic activity needed to guide the economy, there is little rationale as to why governments and banks cannot require the same data on blockchain transactions, which are more reliable, accurate, and quicker to complete and capture

One of the world's top five accounting firms, PwC, reported recently that cryptocurrencies had proven their ability to survive several formidable tests of their legitimacy. They go on to state that the cryptocurrency market remains fragile due to its "dark side"—the exploitation of its benefits by hackers and criminals. PwC advises that the next phase of the cryptocurrency market is mainstream acceptance and stable expansion, in which each of the five key market participants—merchants and consumers, technology developers, investors, financial institutions, and regulators—will play a role.

Whether because mathematical concepts behind cryptocurrencies originated in Asia or because many Asian economies, like Japan's, have more digital than non-digital financial transactions, Asian countries have been very active in creating cryptocurrency volatility. China has aggressively regulated the cryptocurrency impact on its economy, at times halting cryptocurrency payments and at other times allowing them. As early as 2013, Chinese officials prohibited finance and payment companies from buying, selling, quoting prices in, or insuring products linked to Bitcoin after the search engine company, Baidu, opened Bitcoin as a payment method to its 570 million Chinese customers (Hendrickson et al. 2016). Other countries like India, Bolivia, Bangladesh, Thailand, and Vietnam have also banned cryptocurrencies at times. These actions have sent the bitcoin value plummeting.

In less robust national economies like Eastern Europe and northern Africa, cryptocurrency mining operations have emerged. Soviet Georgia has more megawatts of crypto-mining, creating jobs and revenue, than any country other than China. The U.S. Congress has just introduced a joint bill (H.R. 835) that calls for formal policies on alternative technologies citing financial opportunities, growth of the mobile industry, the development of blockchain, cases of identity theft, and transparency of non-fiat currencies. (Young 2016). Now countries like China,

Senegal, Singapore, Tunisia, and Ecuador are issuing their own cryptocurrency, with Estonia, Japan, Palestine, Russia, and Sweden expected to issue a cryptocurrency in the near future (Mason 2017). With so many countries embracing the concept of cryptocurrencies, the cross-border impact is undeniable.

Anticipated Impact on Global Economy

The world has watched as China has manipulated the Bitcoin market to its advantage, first gaining mass Bitcoins through its domination of cryptocurrency mining, then putting restrictions or bans on the currencies, crushing its mining competitors; and now issuing its own Bitcoin. Each time China has made a major policy change on cryptocurrencies, the world value of Bitcoin, in particular, has shifted dramatically.

McKinsey & Co. reported in their first quarter report of this year that, while investors have been fluctuating in their capital contribution to blockchain-related companies, the financial institutions are investing 60% more in their own in-house use of blockchain technology. They further report that central banks are very focused on cryptocurrency issues while regulators are not (McKinsey & Co. 2017). This could well mean that countries like China that are in the forefront of cryptocurrency regulation will affect the economies of other countries as the use of cryptocurrencies rises. Already China's policies have affected the Bitcoin market tremendously, while the U.S. government has had no real impact. Extrapolating data from the Global Cryptocurrency Benchmarking Study (Hileman and Rauchs 2017), if cryptocurrencies continue to follow the growth path of plastic card transactions (which were approximately one-half trillion dollars in 2015), they will undoubtedly have a tremendous impact on global economies.

Ernst & Young dittos this recognition and states:

> In the case of distributed ledger (or blockchain) technology, . . [projects] have already been piloted by ... Barclays, Goldman Sachs and UBS, for transaction settlement and corporate trade finance businesses. The World Economic Forum estimates that four-fifths of the world's commercial banks will have initiated projects using the technology in 2017. With central banks around the world exploring the use of blockchain for creating their own digital currency, . [t]hey will also be forced to address the implications for the wider economy of the return to a (digital) gold standard, and where they sit in regulating money, both old and new (Ernst & Young 2017).

Countries, particularly those with large economies, cannot afford to continue to ignore the cryptocurrency elephant in the room. They do so at their economic peril.

Recommendations

Just three years ago Clark, et al. writing on the economics of information security stated that it would be "long, hard work of integrating the technology into Internet infrastructure and existing institutions" (Clark et al. 2014). Yet, here we are with American banks alone spending $400 million on blockchain technology in 2017. The cryptocurrency market has reached a critical mass of acceptance in the marketplace and new cryptocurrency launches are raising hundreds of millions of dollars in half a year—more quickly than any other category of new company launches. Banks are racing to embrace blockchain technologies not only for its security benefits, but to be able to capitalize on the vast financial potential of serving this emerging marketplace.

Governments allowed the world banks to go unwatched for decades until the major U.S. banks defaulted in 2008 and European banks collapsed in 2009. To prevent this in the future, a mathematician developed the blockchain algorithms of cryptocurrency so that the world economy would have a viable avenue to avoid corrupt banks and governments or manipulation by the few over the many. So how do we best shepherd the emerging cryptocurrency phenomena into our society, our institutions and our lives?

The following recommendations are offered as a starting point to balance the current national currency institutional structures with the principles of blockchain cryptocurrencies:

- Allow small blockchain transactions to go unregulated for certain goods and services.
- Require governments, businesses and institutions that use open source technology to limit its fees, profit, and/or charges therefrom.
- Economically disenfranchised countries, such as Greece, should develop their own integrated cryptocurrency within their national banks.
- Governments should allow its citizens to use more than one currency.
- Require hard or "valued-by-many" assets to back the blockchain for all currencies.
- Governments should be required to use smart contracts, open-ledger technologies to create more transparency and to eliminate fraud, the latter being the most important.
- With technology innovating faster than governments and institutions can regulate, protect, or grasp, we must change the way we promulgate regulations and allow a dynamic input and equal dialogue from all stakeholders and experts.

Conclusion

As societies evolve and technology develops, the types of currencies and institutions that our governments utilize must change. This has never been truer than today. Governments should embrace the benefits of cryptocurrencies: the prevention of fraud, the reduction in costs, extremely reliable transactions, less hackable/more secure financial transaction data and less incentive for politicians, bankers, and power brokers to be corrupt. Equally, governments should regulate the darker side of cryptocurrencies—the facilitation of a financial avenue for illegal transactions, cross-border transactions with unfavored nations, untaxed financial transactions, undisclosed foreign transactions, transactions not computed toward a country's GNP, and breach of cross-border data treaties/laws.

Still, as we move into an increasingly global economy, is there any better way to facilitate nations working together than through an immutable global ledger such as that created by cryptocurrencies? As Accenture points out in an industry report (2015), there are still many questions about how to implement blockchain technologies. However, the time to plan and execute is now. To do this we must decide:

- At what point is a thing valued by independent parties, but not formally recognized by governments, a currency?
- If the blockchain technology is owned by the peoples of the world, does any single government have the right, obligation, or privilege to control it outside of its borders?
- Can and how do governments balance the individual right to privacy with the need for security and the desire to thwart illegal transactions?
- What ethics should be employed in governmental use, control and restrictions on open source technology?
- How do we balance an individual's right to contract with a government's need to tax?
- How do we responsibly transform government, laws, and institutions with the advent of technologies that threaten to make them redundant or massively overemployed?

Cryptocurrencies are the product of an ever-increasing technological and global economy. They are here to stay, so the use of their underlying blockchain technology and regulatory issues need to be addressed now. All facets of society in countries around the world are likely to be impacted by blockchain technology just as they have been by industry from the Industrial Revolution. Cryptocurrencies could catapult us to create better scientific, technological, economic, and political paradigms for living together in society if we seize the moment rather than act in fear of disturbing the status quo.

References

Accenture. (2015). *Blockchain technology: Preparing for change*. Blockchain in the Investment Bank.

Böhme, R. (2015). Bitcoin: Economics, technology and governance. *Journal of Economic Perspectives, 29*(2).

Casey, M. (2014). *BitBeat: New consumer protection laws might cover bitcoin wallets*. Wall St. J. Moneybeat Blog, Retrieved November 13, 2014, from http://blogs.wsj.com/moneybeat/2014/11/13/bitbeat-new-consumer-protection-laws-might-cover-bitcoin-wallets/.

Clark, J., Bonneau, J., Felten, E. W., Kroll, J. A., Miller, A., & Narayanan, A. (2014). On decentralizing prediction markets and order books. In *Workshop on the Economics of Information Security*, State College, PA. Retrieved http://www.econinfosec.org/archive/weis2014/papers/Clark-WEIS2014.pdf.

Ernst & Young GM Ltd. (2017). *The future of money/back to the future: The internet of money*. EYG No. 01602-172Gbl.

Github, Inc. (2017). Retrieved https://github.com/topics/cryptocurrencies.

Gross, P. (2014). A history of virtual currency: Why bitcoins shouldn't surprise you. In: *69th CFA Institute Annual Conference, January 10*.

Han, S., Liu, X., Mao, H., Pu, J., Pedram, A., Horowitz, M. A., Dally, W. J. (2016). EIE: Efficient inference engine on compressed deep neural network. In *Proceedings of the International Symposium on Computer Architecture*. Retrieved https://arxiv.org/pdf/1602.01528v2.pdf.

Heathman, A. (2017). *Move Over Bitcoin, these countries are creating their own digital currencies*, September 27, 2017. Verdict.

Hendrickson, J., Hogan, T., & Luther J. (2016).The political economy of bitcoin. *Economic Inquiry, 54*(2) 925–939, April 2016.

Hileman, G., & Rauchs, M. (2017). *Global currency benchmarking study*. Cambridge Centre for Alternative Finance: University of Cambridge, Judge Business School.

KPMG. (2015). *The Changing World of Money*, January 2015.

KPMG. (2017). *Digital Currencies for a Digital World*.

Mason, J. (2017). *The next cryptocurrency evolution: Countries issue their own digital currency*. Retrieved https://www.fxempire.com/education/article/the-next-cryptocurrency-evolution-countries-issue-their-own-digital-currency-443966.

McKinsey & Company. (2017). *What is blockchain?* Quarterly meeting of the Federal Advisory Committee on Insurance, January 5.

Norry, A. (2017). *The history of the Mt. Gox Hack: Bitcoins biggest heist*. Blockonomi, November 29, 2017.

Obstfeld, M., & Taylor, A. (2017). International monetary relations: Taking finance seriously. *Journal of Economic Perspectives, 31* (3), 3–28, Summer.

Price Waterhouse Cooper. (2015). *Money is no object: Understanding the evolving cryptocurrency market*, August 2015.

Raj, S. (2018). *Data story: As the price of bitcoin soared, its dominance in the market plummeted by 57% in 2017*. Retrieved January 2018, from http://www.moneycontrol.com/news/business/markets/data-story-as-the-price-of-bitcoin-soared-its-dominance-in-the-market-plummeted-by-57-in-2017-2481773.html.

Wagner, A. (2014). *Digital vs. virtual currencies*, BitCoin magazine, 22 August 2014.

Yap, B. (2016). *Regulation vs. innovation in the internet of value*. Retrieved June 19, 2016, from www.bitlegal.net.

Young, J. (2016). *S. House resolution calls for national bitcoin and blockchain policy*. Retrieved July 21, 2016, from BTCManage.com.

Part VI
Messages and Recommendations

Message from Ms. Irina Bokova, Director-General of UNESCO on the Occasion of the 9th Olympiad of the Mind Chania, Greece, 14–17 September 2017

Epimenidis Haidemenakis

Excellences, Ladies and Gentlemen, Dear Participants

The theme of the *9th Olympiad of the Mind* stands at the heart of all UNESCO's work. 'Learning to live together' is the red thread that ties together all of UNESCO's action to strengthen the foundations for peace and sustainable development through cooperation in education, culture, the sciences, culture and communication. This cross-sectoral work echoes the spirit of the STEPS concept, which seeks to join Science, Technology, Economics, with Politics and Society.

This has never been so important at time when all societies are transforming and increasingly diverse. Learning to live together is both the goal and the path, based on respect for the inherent rights and dignity of every woman and man. All cultures are different—this is a boon for creativity and innovation, exchange and cooperation—but we stand together as a single humanity, sharing freedoms and aspirations, a past and a future. Learning this must start as early as possible, on the benches of schools, through education for peace and human rights, education for global citizenship, and taken forward through lifelong learning for all.

This is the spirit of the *2030 Agenda for Sustainable Development*, and specifically its Sustainable Development Goal 4 on education. Education is a development multiplier for achieving all of the Sustainable Development Goals, to leave no one behind, to deepen sustainability, and to build more inclusive, peaceful and just societies.

The world faces many challenges today, from conflicts to climate change. These are not technical—their solutions require knowledge, values, understanding and imagination. UNESCO's recent *World Humanities Conference*, held in Belgium in August 2017, expressed a vision of how academic scholarship, social mobilization

E. Haidemenakis (✉)
Olympiad of the Mind and the International S.T.E.P.S. Foundation, New York, USA
e-mail: steps.peace@gmail.com

© Springer International Publishing AG, part of Springer Nature 2019
J. A. S. Kelso (ed.), *Learning To Live Together: Promoting Social Harmony*,
https://doi.org/10.1007/978-3-319-90659-1_24

and public debate can work together to open new spaces for the future of humanity. I count on the *9th Olympiad of the Mind* to enrich this vision ever more, and I thank you all in advance for your commitment to the goals we share.

Irina Bokova

The Relevance of the Ancient Greek Texts

Alexandros P. Mallias

Ladies and gentlemen,

I am most humbled by the invitation to speak among such distinguished servants of episteme, of logos and logiki (logic), and I thank Dr. Epimenidis D. Haimenedakis for extending that invitation.

The mere fact that I am surrounded by Nobel Prize recipients will help reign me in and prevent me from expounding at length on a topic that I love, that is, the "Relevance of the Greek Classics today".

I will try to develop within my brief remarks the issue of the lack of symmetry, harmony and geometry in the 21st century world.

Ancient Greeks recognized that man is part of a greater whole, and it is obvious today that the safety of the world rests upon the realization that our fates are intertwined and interwoven; we are all part of a greater whole, which needs balance and equilibrium. This balance requires the blend of harmony, symmetry, geometry and a sense of measure (metron), qualities that the ancient Greeks understood better than anyone.

These qualities are explicit and mirrored in classical Greek sculpture. Ancient Greek statues and temples are all on a human scale, something which shows a profound understanding of man's proportionate relationship to nature and the cosmos. You only have look to the Parthenon, a structure which embodies these characteristics, regrettably disrupted by the fact that the Parthenon marbles are in the British Museum.

I find that many of the problems and challenges we face today are precisely due to the fact that these qualities arc missing.

Editor's Note. This speech was given at the Eighth Olympiad of the Mind hosted by The National Academies, Washington, DC. We were fortunate that Ambassador Mallios also participated actively at the Ninth Olympiad, in Chania, Crete.

A. P. Mallias (✉)
The National Academies, Washington, D.C, WA, USA

© Springer International Publishing AG, part of Springer Nature 2019
J. A. S. Kelso (ed.), *Learning To Live Together: Promoting Social Harmony*,
https://doi.org/10.1007/978-3-319-90659-1_25

1. And so, I often find solace and counsel, if you will, in the ancient Greek classics, as they negotiate ways to maintain this balance and harmony in relationship to the whole:

(A) On Geometry—ΓΕΩΜΕΤΡΙΑ

Clearly, the concept of Geometry includes the sub-concepts of "geo" (earth) and "metron" (measurement). It is not very difficult to argue that today in international relations at the global level the essence of metron is lacking.

(B) On Harmony—(ΑΡΜΟΝΙΑ)

Harmony presupposes a fine-tuned equilibrium and a proportionate relationship of all components of all our regulatory systems, including our ecosystem.

Climate change and global warming are but blatant examples of this lack of harmony.

In the 21st century, we have introduced the concept of asymmetrical threats. It is clear to me that the environment and global climate are victims of asymmetrical threats of our own making.

The Greek classics tackle this issue as well. I was amazed to read Plato's bemoaning the abuse of the environment, when in his dialogue "Critias", he talks about the Attic land:

> For some mountains, which today will only support bees produced not so long ago, trees which when cut provided roof beams for huge buildings whose roofs are still standing. And there were a lot of tall cultivated trees which bore unlimited quantities of fodder for beasts. The soil benefited from an annual rainfall which did not run to waste from the bare earth as it does today, but was absorbed in large quantities and stored in retentive layers of clay, so that what was drunk down by the higher regions flowed downwards into the valleys and appeared everywhere in a multitude of rivers and springs. And the shrines which still survive at these former springs are proof of the truth of our present account of the country.

C. On Symmetry (ΣΥΜΜΕΤΡΙΑ)

The lack of symmetry, such an important concept in the Greek classics, manifests itself in at least two main ways today:

First, the lack of symmetry manifests itself in the gap between rich and poor.

Aggregate wealth estimates provided by the World Bank demonstrate that the European countries, along with the United States, and Japan, dominate the top 10 wealthiest countries/nations. The 10 poorest countries at the global level are in Sub-Saharan Africa.

It is with the rise of the Greek city-States that we see a civilization concerned with the delicate balance between food supply and population. Ancient Athens was especially troubled by demographic pressures. Thus the ancient Greek philosophers, particularly Plato and Aristotle, were sensitive to the relationship between population and resources when contemplating the ideal size for a city-state of their day.

A second phenomenon resulting from this gap, and allow me to be a bit of a heretic here, is that of human trafficking. In ancient and not so ancient times, we know there were slaves; there were the dominant and the dominated. Today, albeit within a different context and perhaps different form, we have generated the phenomenon of a modem form of slavery, that of human trafficking.

2. LIVING TOGETHER IN DIVERSITY

A basic characteristic of Athenian democracy was the concept of co-existing, of living together in diversity.

Today, both in the United States and in a number of European countries, a critical phenomenon that has entered the domestic political agenda is that of **immigration**.

And given these values and concepts of ancient Athens, it did not surprise me to discover that Martin Luther King himself was interested in the classics and so freely referred to them in his many great speeches, including the most famous "To the Mountaintop"

3. THUCYDIDES

In understanding the miscalculations and blunders in political decision-making in the 21st century, but also looking back to the 20th century, there is no better handbook than Thucydides' "History of the Peloponnesian Wars", written some 2,500 years ago.

A. PERSUASION RATHER THAN FORCE and HOW TO TREAT OUR ALLIES:

In the 21st century, the concept of the Coalition of the Willing has emerged. In his famous speech to Sparta's allies, on the merits of going to war, Archidamus, the Spartan King, says:

B. ALLIES:-COALITION OF THE WILLING:

"What is most important is that we shall have all mankind as our allies—not because they have been forced, but rather persuaded, to join with us; who will not welcome our friendship because of our power, but who will be disposed towards us as allies and friends.

We Must be willing to treat our allies as we would our friends and not to grant them ... things ... only in words, ... and not exercise our leadership as masters but as helpers."

So we shall not lack allies to help us ... but shall find many ready and willing to join their forces to our own. For what city or what men will not be eager to share our friendship and our alliance when they see that we are at once the most just and the most powerful of peoples?

C. ON PREPARATIONS FOR WAR: "If we undertake the war... we would by hastening its commencement only delay its conclusion."

He continues, "In practice we always base our preparations against an enemy on the assumption that his plans are sound—indeed, it is right to rest our hopes not on a belief in his blunders, but on the soundness of our provisions."

"We must not rest our hopes of safety upon the blunders of our enemies but upon our own management of affairs and upon our own judgment. For the good fortune which results to us from their stupidity might perhaps cease or change to the opposite, whereas that which comes about because of our own efforts will be more permanent and enduring."

CONCLUSION

The return/revival of the classics in the 21st century as a basic tool of analysis, of understanding the sound criteria for decision making in politics, geostrategy is to myself imposed in our permanent search for applying the themelion (fundamental principles) of symmetry, harmony and geometry, which is the Aristotelian metron.

Recommendations from the Ninth Olympiad of the Mind: Learning to Live Together

The International S.T.E.P.S. Foundation

The Olympiad of the Mind is like a large think tank. The speakers are experts from countries around the globe who are thought leaders, decision makers and policy makers in their own respective countries and regions. Every speaker was asked to give at least two recommendations at the end of their paper for "policy" and action. As we cannot effectively address each and every recommendation, the participants of the 9th Olympiad have integrated the spirit, objectives and goals of over seventy recommendations into the following twelve:

1. Countries must respond as one global unit to global problems (e.g. reducing climate change) by **connecting knowledge to action** as soon as possible.
2. Every country should **enhance support of scientific research** and **technology** in all fields (including bio-, nano- and health) that promote global harmony with the potential **to improve quality of life**.
3. Developing countries should promote and develop the **political will for funding** to create a **critical mass of specialists** in all areas of science, technology, innovation and financial systems.
4. Promote **diversity in education** and **variability in our lives**.
5. Make **a paradigm change** whereby **education** expands beyond raising economic agents to **raising ethical agents** who embrace the politics of difference and actively **respond to global issues**.
6. **Free high-quality education** at every level for all.
7. In addition to mathematics, science and technology, education needs to ensure that each student has a variety of opportunities to learn about **humanities** and **global issues through direct and indirect experience** (e.g., ecoliteracy, non-violence, etc.).

The International S.T.E.P.S. Foundation (✉)
Science, Technology, Economics and Politics for Society, Chania, Greece
e-mail: jascott.kelso@gmail.com; klight777@msn.com
URL: https://www.internationalstepsfoundation.org

© Springer International Publishing AG, part of Springer Nature 2019
J. A. S. Kelso (ed.), *Learning To Live Together: Promoting Social Harmony*,
https://doi.org/10.1007/978-3-319-90659-1_26

8. Provide **financial support** to organizations that promote science, technology, innovation and economics in the interest of peace and society (e.g. The **International STEPS Foundation** and **its goals**).

9. Create and support **sustainable "smart" cities** and communities.

10. Support an understanding for the necessity to **reduce fossil fuels** as an energy source.

11. Recognize **individual variability**, support **peaceful** and **diversified cultures**, and promote **gender equality** within **all societies**.

12. **Support the U.N.'s seventeen Sustainable Development Goals**, for each person to have clean water, food, energy, high quality education, employment and transportation to it, as well as good healthcare.

WE encourage ALL—individuals and institutions alike—to implement these *Recommendations* within your sphere of influence whether that be within your family, your school, your work, your government or the world at large. WE welcome your feedback.

PGMO 06/08/2018